Higher Biology

15

How to Pass

SECOND EDITION

HIGHER

Biology

Billy Dickson and
Graham Moffat

HODDER GIBSON
AN HACHETTE UK COMPANY

Orders: please contact Bookpoint Ltd, 130 Park Drive, Milton Park, Abingdon, Oxon OX14 4SE. Telephone: (44) 01235 827827. Fax: (44) 01235 400454. Email education@bookpoint.co.uk Lines are open from 9 a.m. to 5 p.m., Monday to Friday, with a 24-hour message answering service. Visit our website at www.hoddereducation.co.uk. If you have queries or questions that aren't about an order, you can contact us at hoddergibson@hodder.co.uk

First published in 2014 © Graham Moffat and Billy Dickson
This second edition published in 2019 by
Hodder Gibson, an imprint of Hodder Education
An Hachette UK Company
211 St Vincent Street
Glasgow, G2 5QY

Impression number 5 4 3

Year 2023 2022 2021 2020

Cover photo © arafel – stock.adobe.com
Illustrations by Aptara, Inc.
Typeset in CronosPro-Lt 13/15 pts by Aptara, Inc.
Printed in India
A catalogue record for this title is available from the British Library.
ISBN: 978 1 5104 5222 0

Contents

4 Skills of scientific inquiry

Introduction

General introduction

Welcome to *How To Pass Higher Biology*

The fact that you have opened this book and are reading it shows that you want to pass your SQA Higher Biology Course. This is excellent because that type of attitude is needed in order to pass and pass well. It also shows you are getting down to the revision that is *essential* to pass and get the best grade possible.

The idea behind the book is to help you to pass, and if you are already on track to pass, it can help improve your grade. It can help boost a D into a C, a C into a B or a B into an A. It cannot do the work for you, but it can guide you in how best to use your limited time.

In producing this book we have assumed that you have followed an SQA Higher Biology Course at school or college this year and that you have probably, but not necessarily, already studied National 5 Biology.

We recommend that you download and print a copy of the Higher Biology Course Specification from the SQA website at www.sqa.org.uk.

You should note that, in your exam, only the material included in the Key Areas and Depth of Knowledge required columns on pages 33–87 of the Course Specification can be examined. Skills of scientific inquiry, described on page 5 of the Course Specification, are also examined. You should get copies of any specimen or past papers that are available on the SQA website.

Course assessment outline

The components of the Higher Biology assessment system are summarised in the table below.

Higher Biology	Assessment	Who does the assessing?
Course (graded A–D)	**Components 1 and 2: Examination** (worth 80% of the grade) 25 multiple choice items worth 1 mark each and 95 marks allocated to short and longer answer questions	Marked by SQA out of 120 marks
	Component 3: Assignment (worth 20% of the grade)	Marked by SQA out of 20 marks (but scaled to 30)

Components 1 and 2: Course examination

(120 marks)

The Higher examination consists of two papers.

- **Paper 1** (component 1) contains 25 multiple choice questions for 1 mark each.
- **Paper 2** (component 2) contains a mixture of structured short answer questions and extended response questions for a total of 95 marks. Some questions will have an element of choice within them.

The majority of the marks test demonstration and application of knowledge, with an emphasis on the application of knowledge. The remainder test the application of scientific inquiry skills.

The course examination is marked by SQA and contributes 80% to the overall grade for the Course.

A list of the knowledge and skills required for your exam is summarised in the table below.

Knowledge and Skills	Meaning
Demonstrating knowledge	Demonstrating knowledge and understanding of Key Areas of biology
Applying knowledge	Applying knowledge of Key Areas of biology to new situations, interpreting information and solving problems
Planning	Planning and designing experiments to test given hypotheses
Selecting	Selecting information from a variety of sources
Presenting	Presenting information appropriately in a variety of forms
Processing	Processing information using calculations and units where appropriate
Predicting	Making predictions and generalisations based on evidence
Concluding	Drawing valid conclusions and giving explanations supported by evidence
Evaluating	Suggesting improvements to experiments and investigations

The knowledge content is detailed in the tables on pages 33–74 of the Course Specification and you should note that both the Key Area and Depth of knowledge required columns can be examined.

You should be familiar with the Apparatus and techniques detailed on page 75 of the Course Specification.

Component 3: Assignment

(20 marks)

The Assignment is a task which is based on some research that you have carried out in class time. The research must involve experimental work which allows measurements to be made. Candidates must also gather data/information from the internet, books or journals. The research will be supervised by your teacher and you will have to write up the investigation in the form of a Report during a controlled assessment. During the controlled assessment you will have access to selected material and notes, but you cannot use a draft copy of any part of your Assignment Report.

The Assignment has two stages:

1 a research stage
2 a report stage.

The Assignment Report

The Report is marked out of 20 marks, allocated as shown in the table below. There are 16 marks for skills and 4 marks for the demonstration and application of biological knowledge.

Section of your Report	Marks
Stating your aim	1
Underlying biology	4
Data collection and handling	5
Graphical presentation	4
Analysis	1
Conclusion	1
Evaluation	3
Structure	1
Total	**20**

The Assignment Report is marked by SQA. Although the Report is marked out of 20 marks, SQA will scale the mark up to 30, so that it makes up 20% of your overall course assessment and grading.

About this book

We have tried to keep the language used in this book simple and easy to understand and we have used the language of the Higher Biology Course Specification. This is the language used in the exam papers.

We suggest that you use this book throughout your Course. Use it at the end of each Key Area covered in class, at the end of each Area of biology in preparation for end-of-Area class assessment, before your preliminary examination and, finally, to revise the whole Course in the lead up to your final examination.

There is a grid on page x that you can use to record and evaluate your progress as you finish each Area of biology.

The Course content section is split into three sections, which cover the three Areas of Higher Biology. Each Area is divided into Key Areas, each of which covers a Course Specification Key Area. The Key Areas have four main features: Key points, Summary notes, Key words and Questions.

Key points 

Key points boxes list the pieces of knowledge that you need to demonstrate and apply in each Key Area. You can use these boxes to monitor your learning. You could use a traffic-light system to code each box – green for fully understood, orange for nearly there and red for needing to cover again.

The words in **bold** within Key points are vital and each is defined in the Key words section at the end of the chapter and also in the Glossary at the back of the book.

Summary notes

These give a summary of the knowledge required in each Key Area. You must read these carefully. You could use a highlighter pen to emphasise certain words or phrases and you might want to add your own notes in the margin in pencil or on sticky notes.

There are diagrams to illustrate many of the key learning ideas.
The summary notes often contain Hints & tips boxes like that shown here.

T

In addition, the symbol shown on the left indicates that an important technique, which you need to know about for your exam, is described within the notes. There is always a Key links box beside the symbol which links to a question on the technique in the Skills of scientific inquiry chapter on page 146.

Hints & tips

Where we offer a tip to help learning it is boxed like this. These tips can be very general or can be specific to the content of the Key Area. The tips are suggestions – don't feel you need to use them all.

Key words

Important words are listed at the end of each chapter along with their definitions. These are the crucial terms you need to know to do well in Higher Biology. After completing your work on a chapter, it is a good idea to make a set of flash cards using these words.

Questions 

A set of short answer and longer answer questions is found at the end of each chapter. These allow you to practise the demonstration of knowledge from the Key Area, and include a technique question and symbol as appropriate.

Standard answers are provided at the end of each set of chapters and it is recommended that you mark and correct your own work.

Key links 

These boxes alert you to topics that are linked to other Areas in the Course, and direct you to sections of the book where you can read or do more.

Practice course assessment

We have included a Practice course assessment linked to each Area. These can give you an idea of your overall progress in the course. We have designed these Practice assessments to be like mini exam papers with multiple choice, structured and extended response items.

The questions are intended to replicate those you will meet in the course exam. They allow you to judge how you are doing overall. Some questions test knowledge and its application, while others test Skills of scientific inquiry. Question types are provided in roughly the same proportion as in your final exam.

Give yourself a maximum of 75 minutes to complete each test.

Mark your own work using the answers provided at the end of each Area. You could grade your work as you go along to give you an idea of how well you are doing in the course. The table below shows a suggested grading system:

Mark out of 50	Grade
20–24 marks	D
25–30 marks	C
31–35 marks	B
36+ marks	A

Skills of scientific inquiry: three approaches

The section on pages 146 to 168 covers the Skills of scientific inquiry and includes questions to test these. We have given three different approaches to revising these science skills and recommend that you use all three:

4.1 gives an example of a scientific investigation and breaks it down into its component skills. There are questions on each skill area.
4.2 goes through the skills one by one and gives you some exam-style questions covering the apparatus, techniques and skills you should know about.
4.3 provides sets of hints and tips on answering science skills exam questions.

Your Assignment

We give an introduction to the Assignment, some suggestions for suitable topics and some information, with hints, to help you complete the task. The table on page 172 has a checklist of points you need to cover in your Report.

Your exam

We give some hints on approaches to your final exams in general, as well as more specific tips for your Higher Biology exam.

Glossary

Here we have brought together the Key words that occur in the Course Specification for Higher Biology and their definitions in the context of the Key Area in which they first appear in the book. You could use the Glossary to make flash cards. A flash card has the term on one side and the definition on the other. Get together with a friend and use these cards to test each other.

Answers

Answers are provided for all of the questions in this book. These are intended to replicate SQA standard answers but we have tried to keep the answers short, and any instructions simple, to make them easier to use – there will be other acceptable answers.

Record of progress and self-evaluation

Use the grid below to record and evaluate your progress as you finish each of the Key Areas, and once you have completed and marked the Practice course assessment.

Key Area			Progress indicator				Key Area feedback 1 – OK 2 – more work 3 – help needed
			Key points traffic-lighted	Summary notes read and understood	Questions tried and self-marked	Key words flash cards made	
DNA and the genome	1.1	Structure of DNA					
	1.2	Replication of DNA					
	1.3	Gene expression					
	1.4	Cellular differentiation					
	1.5 & 1.6	Structure of the genome and mutation					
	1.7	Evolution					
	1.8	Genomic sequencing					
	Practice course assessment		Paper 1	/10	Paper 2	/40	Total /50
Metabolism and survival	2.1a	Metabolic pathways					
	2.1b	Control of metabolic pathways					
	2.2	Cellular respiration					
	2.3	Metabolic rate					
	2.4	Metabolism in conformers and regulators					
	2.5	Metabolism and adverse conditions					
	2.6	Environmental control of metabolism					
	2.7	Genetic control of metabolism					
	Practice course assessment		Paper 1	/10	Paper 2	/40	Total /50
Sustainability and interdependence	3.1a	Food supply					
	3.1b	Photosynthesis					
	3.2	Plant and animal breeding					
	3.3 & 3.4	Crop protection and animal welfare					
	3.5 & 3.6	Symbiosis and social behaviour					
	3.7	Components of biodiversity					
	3.8	Threats to biodiversity					
	Practice course assessment		Paper 1	/10	Paper 2	/40	Total /50

Area 1 DNA and the genome

Key Area 1.1
Structure of DNA

Key points !

1 Genetic information is inherited. ☐
2 **DNA** is a substance that encodes the genetic information of heredity in a chemical language. ☐
3 DNA is a very long double-stranded molecule in the shape of a **double helix**. ☐
4 Each strand is made up from chemical units called **nucleotides**. ☐
5 A nucleotide is made up of three parts: a **deoxyribose** sugar, a **phosphate** and a **base**. ☐
6 Deoxyribose molecules have five carbon atoms, which are numbered 1 to 5. ☐
7 The phosphate of one nucleotide is joined to carbon 5 (5') of its sugar and linked to carbon 3 (3') of the sugar of the next nucleotide in the strand to form a **3'–5' sugar–phosphate backbone**. ☐
8 There are four different bases called **adenine (A)**, **guanine (G)**, **thymine (T)** and **cytosine (C)**. ☐
9 The nucleotides of one strand of DNA are linked to the nucleotides on the second strand through their bases – the bases form pairs that join the strands together. ☐
10 Bases pair in a complementary way – adenine always pairs with thymine and guanine always pairs with cytosine. ☐
11 Base pairs are held together by **hydrogen bonds**. ☐
12 Each strand has a sugar–phosphate backbone with a 3' end that starts with a deoxyribose molecule and a 5' end that finishes with a phosphate. ☐
13 The two strands of a DNA molecule run in opposite directions and are said to be **antiparallel** to each other. ☐
14 The base sequence of DNA forms the genetic code. ☐
15 DNA is organised differently in different types of organisms. ☐
16 Cells can be divided into two groups based on how they organise their DNA – eukaryotes and prokaryotes. ☐
17 **Prokaryotic** cells do not have a distinct nucleus. They have a single, circular **chromosome** and smaller circular **plasmids**. ☐
18 **Eukaryotic** cells have a nucleus and membrane-bound organelles such as mitochondria and chloroplasts. ☐
19 They all have linear chromosomes in the nucleus, which are tightly coiled and packaged with associated proteins. ☐
20 The associated **proteins** are called **histones**. ☐
21 Eukaryotic cells also contain circular chromosomes in their **mitochondria** and **chloroplasts**. ☐
22 **Yeast** is a special example of a eukaryote as it also has plasmids. ☐

Summary notes

Function of DNA

Genetic information is coded into the chemical language of DNA (deoxyribonucleic acid). This genetic information gives cells the ability to synthesise specific proteins that determine the cell's structure and allow it to control metabolism. Copies of a cell's genetic information are inherited by daughter cells when it divides.

Structure of DNA

Each DNA molecule is very long and has two strands coiled into the shape of a double helix. Each strand of the double helix is made up from nucleotides. Figure 1.1 shows a single DNA nucleotide made up of a deoxyribose sugar to which a phosphate group and a base are attached. The carbon atoms of the deoxyribose sugar are numbered from 1 to 5, as shown in the diagram.

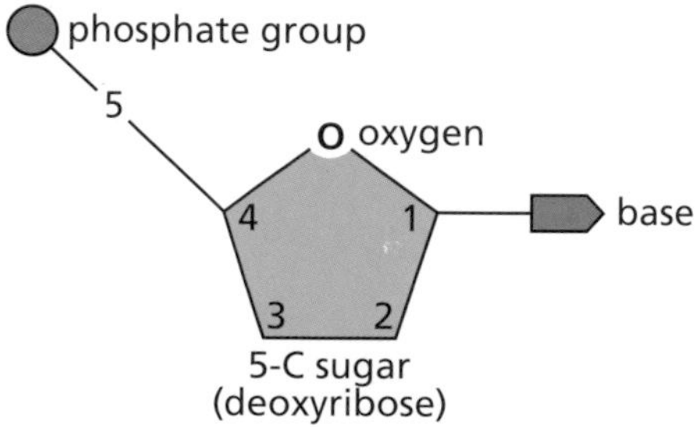

Figure 1.1 One nucleotide of DNA with the carbon atoms of the deoxyribose sugar numbered

Nucleotides are linked by their deoxyribose sugars and phosphates to form a strand with a sugar–phosphate backbone, as shown in Figure 1.2.

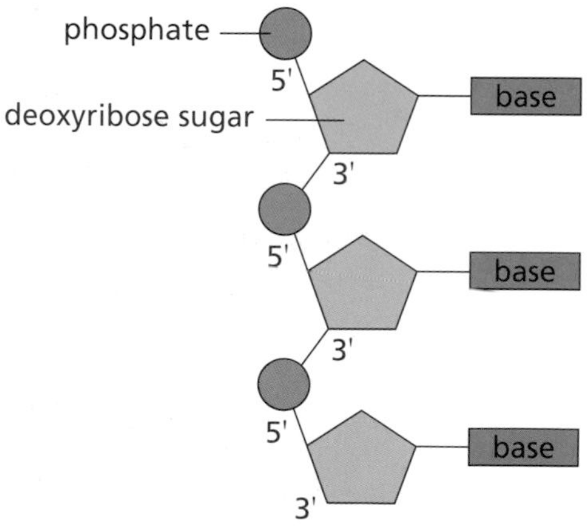

Figure 1.2 Short strand of DNA showing three nucleotides linked by a 3'–5' sugar–phosphate backbone

Two strands are connected by hydrogen bonding between complementary pairs of bases. The base adenine (A) always pairs with thymine (T) and guanine (G) always pairs with cytosine (C) making the two strands complementary to each other, as shown in Figure 1.3. Note that the strands run in opposite directions (antiparallel) depending on the bonding through the carbon atoms of the sugar–phosphate backbone. One strand has deoxyribose (3') at one end of the molecule, but its complementary strand has a phosphate group (5') at the same end of the molecule.

Hints & tips

The DNA strands are a bit like lanes of traffic on a road – they are essentially the same but run in opposite directions. This is what is meant by the term antiparallel.

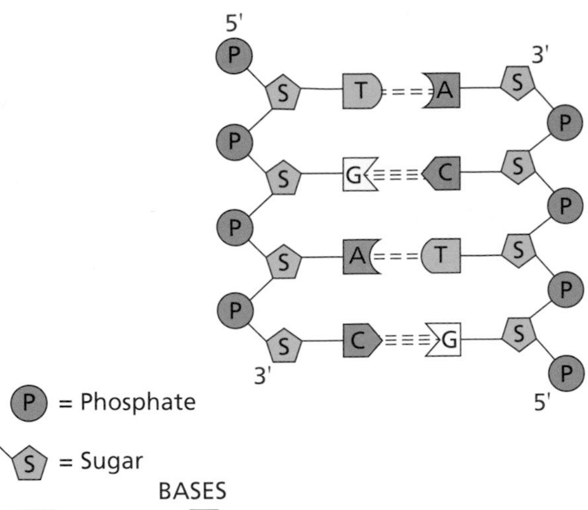

Figure 1.3 Short double strand of DNA showing complementary base pairing and its antiparallel structure

Hints & tips

There are a number of features of DNA molecules that you should note for your exam:

- ✓ *double helix shape*
- ✓ *sugar–phosphate backbones*
- ✓ *antiparallel strands*
- ✓ *hydrogen bonds linking strands*
- ✓ *complementary base pairing rules applied to nucleotides.*

Figure 1.4 summarises the structural features of a DNA molecule.

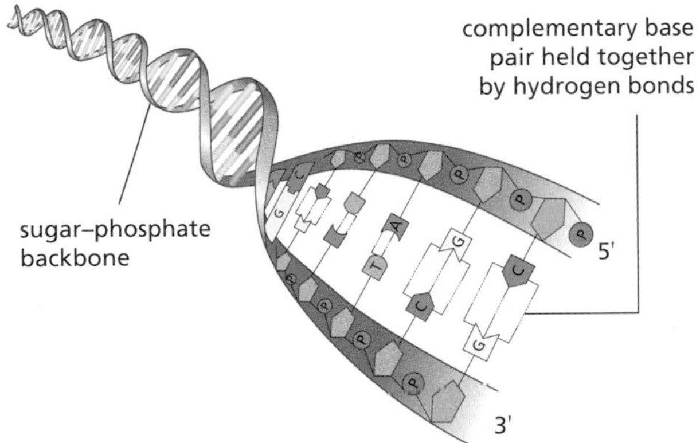

Figure 1.4 DNA showing double helix, sugar–phosphate backbones, complementary bases pairing and the antiparallel strands

Organisation of DNA in cells

DNA is organised differently in the cells of organisms from different domains of life. In the domains Archaea and bacteria the cells are prokaryotic but in the domain Eukarya the cells are eukaryotic.

Prokaryotic cells

In prokaryotic cells such as bacteria there is no distinct nucleus and the DNA is organised into circular chromosomes and plasmids, as shown in Figure 1.5.

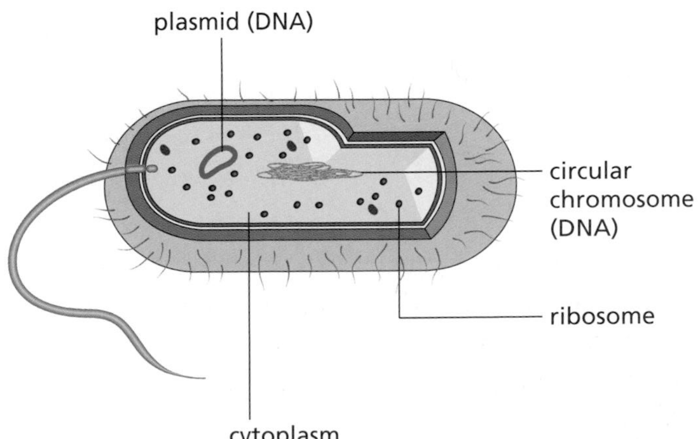

Figure 1.5 Main structures in a prokaryotic bacterial cell, showing circular chromosome and plasmids

Key links

There is more about domains of life in Key Area 1.8 (page 32).

Key links

There is more about prokaryotic cells and DNA in Key Area 2.7 (page 84).

Eukaryotic cells

In the nuclei of eukaryotic cells, the DNA is found tightly coiled into linear chromosomes and associated with proteins called histones that help to keep the DNA strands untangled, as shown in Figure 1.6.

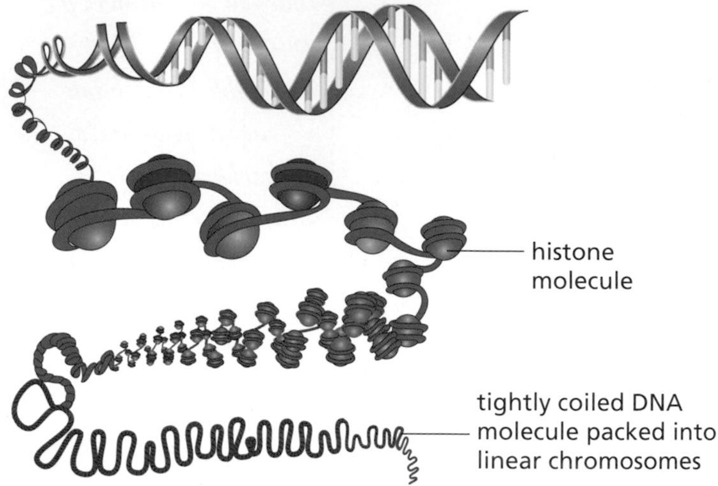

Figure 1.6 Organisation of DNA in eukaryote chromosomes

Eukaryotes also have small circular chromosomes within their mitochondria and within their chloroplasts in the case of plant cells, as shown in Figure 1.7.

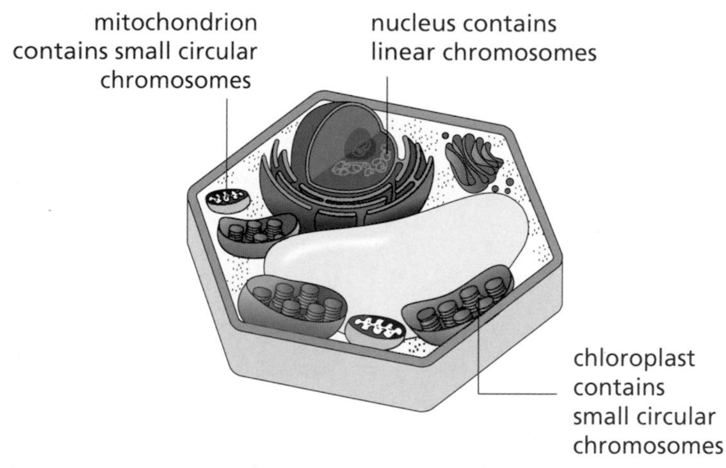

Figure 1.7 Organisation of DNA in a eukaryotic plant cell

Yeasts are special eukaryotes because, although their DNA is also tightly coiled into linear chromosomes within the nucleus, yeast species also have plasmids in their cytoplasm. The following table provides a summary of the organisation of DNA in cells.

Organism	Cell type	Organisation of DNA		
		Linear chromosomes	Circular chromosomes	Plasmids
Bacteria	Prokaryotic	Not present	In cytoplasm	Yes
Animals	Eukaryotic	Within nucleus	Within mitochondria	No
Green plants	Eukaryotic	Within nucleus	Within mitochondria and chloroplasts	No
Yeast	Eukaryotic	Within nucleus	Within mitochondria	Yes

Key words

3'–5' – refers to the direction of a DNA strand with deoxyribose at the 3' end and phosphate at the 5' end
Adenine (A) – base that pairs with thymine in DNA (and uracil in RNA)
Antiparallel – parallel strands of DNA, running in opposite directions
Base – coding component of a nucleotide
Chloroplast – organelle in which the chemical reactions of photosynthesis occur
Chromosome – structure that contains the genetic material of an organism encoded into DNA
Cytosine (C) – base that pairs with guanine in DNA
Deoxyribose – pentose sugar that is a component of a DNA nucleotide
DNA – deoxyribonucleic acid; a molecule that holds the genetic code in living organisms
Double helix – three-dimensional shape of a DNA molecule
Eukaryotic – refers to a domain of life characterised by cells with a discrete nucleus
Guanine (G) – base that pairs with cytosine in DNA
Histones – proteins with which DNA is associated in linear chromosomes
Hydrogen bond – weak chemical link joining complementary base pairs in DNA
Mitochondria – cell organelles in which the aerobic stages of respiration occur (*sing.* mitochondrion)
Nucleotide – component of DNA consisting of a deoxyribose sugar, a phosphate group and a base
Phosphate – component of a DNA nucleotide that forms part of the sugar–phosphate backbone.
Plasmid – circular loop of genetic material found in prokaryotic organisms and some yeasts
Prokaryotic – refers to the domains of life characterised by cells with no discrete nucleus
Protein – large molecule made up from a chain of amino acids linked by peptide bonds
Sugar–phosphate backbone – strongly bonded strand of DNA
Thymine (T) – base that pairs with adenine in DNA
Yeast – a special eukaryote which contains plasmids

Questions ?

Short answer (1 or 2 marks)

1 DNA is a complex double-stranded molecule made up from nucleotide units.
a) Describe the shape of a DNA molecule. (1)
b) Describe how the two strands of DNA are held together. (2)
c) Name the **three** components that make up a nucleotide. (2)
d) Explain what is meant by the following terms as applied to DNA structure:
(i) complementary (2)
(ii) antiparallel. (2)

2 Suggest why yeasts are described as special eukaryotes. (2)

Longer answer (3–10 marks)

3 Cells can be classified as prokaryotic or eukaryotic. Describe the organisation and distribution of DNA in a prokaryotic bacterium and a eukaryotic plant cell. (4)

4 Describe the function of DNA and give an account of the structure of a DNA molecule. (7)

Answers are on page 36.

Key Area 1.2

Replication of DNA

Key points

1. **Replication** is the process by which DNA molecules direct the synthesis of identical copies of themselves. ☐
2. Prior to cell division, DNA is replicated by the enzyme **DNA polymerase**. ☐
3. DNA polymerase needs **primers** to start replication. ☐
4. A primer is a short complementary strand of nucleotides which binds to the 3' end of the **DNA template strand** allowing polymerase to add DNA nucleotides. ☐
5. DNA polymerase adds DNA nucleotides, using complementary base pairing, to the deoxyribose (3') end of the new DNA strand which is forming. ☐
6. DNA is unwound and hydrogen bonds between complementary bases are broken to form two template strands. ☐
7. DNA polymerase can only add DNA nucleotides in one direction from its 3' end towards its 5' end. This results in continuous replication of the **leading strand** while the **lagging strand** is replicated in fragments. ☐
8. Fragments of DNA are joined together by the enzyme **ligase**. ☐
9. The **polymerase chain reaction (PCR)** is a laboratory technique for the amplification of DNA. ☐
10. PCR amplifies DNA using complementary primers for specific target sequences. ☐
11. In PCR, primers are short strands of nucleotides which are complementary to specific target sequences at the two 3' ends of the region of DNA to be amplified. ☐
12. Repeated cycles of heating and cooling amplify the target region of DNA. A cycle of PCR doubles the number of copies of a region of DNA. ☐
13. DNA is heated to between 92 and 98 °C to separate the strands. It is then cooled to between 50 and 65 °C to allow primers to bind to target sequences. ☐
14. It is then heated to between 70 and 80 °C for **heat-tolerant DNA polymerase** to replicate the region of DNA. ☐
15. PCR has a variety of practical applications. PCR can amplify DNA to help solve crimes, settle paternity disputes and diagnose genetic disorders. ☐
16. Macromolecules such as fragments of DNA from a source can be separated by **gel electrophoresis**. ☐

Summary notes

Replication of DNA

DNA is the hereditary material of cells. It can make identical copies of itself by a process called replication. DNA replicates prior to cell division and copies are passed to daughter cells.

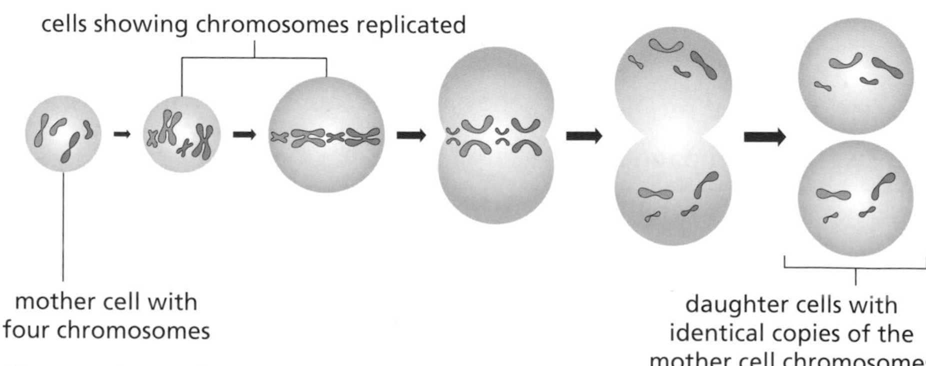

Figure 1.8 Stages of mitosis

Figure 1.8 shows a cell with four linear chromosomes before and after their DNA has replicated and how the chromosomes then move to form daughter cells.

Stages in replication of DNA

The double helix of DNA is unwound by an enzyme and the hydrogen bonds that connect the two strands are unzipped. The unwinding and unzipping form a replication fork.

Primers are short complementary strands of nucleotides that allow DNA polymerase to bind. A primer joins the 3' end of the 3'–5' leading template strand and DNA polymerase adds free complementary DNA nucleotides to synthesise a complementary strand continuously.

DNA polymerase can only add nucleotides in the 3' to 5' direction, so on the lagging strand primers are added one by one into the replication fork as it widens. DNA nucleotides are added in fragments. These fragments are then joined by DNA ligase to form a complete complementary strand. The process requires energy, which is supplied by ATP produced by the cell's respiration. The replication process is summarised in Figure 1.9.

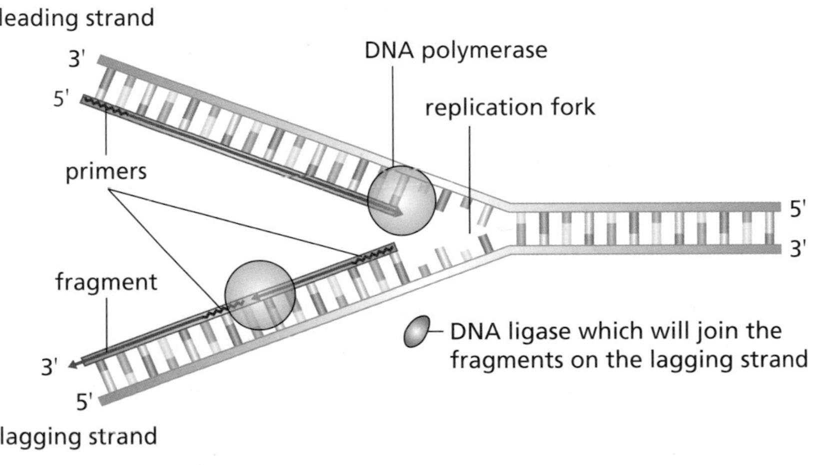

Figure 1.9 Replication of DNA

Importance of DNA replication

DNA replication is important because it ensures that identical copies of the genetic information of a species are passed on from cell to cell and from generation to generation, as shown in Figure 1.10. Each daughter cell has all the genetic information needed to produce the proteins to carry out its functions.

Key links 

There is more about ATP production in Key Area 2.2 (page 59).

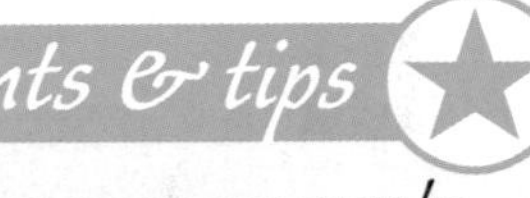

In your exam you may be asked to state the requirements for DNA replication – these are DNA templates, free DNA nucleotides, primers, DNA polymerase and a source of energy (ATP).

Key links 

There is more about the synthesis of proteins in Key Area 1.3 (page 11).

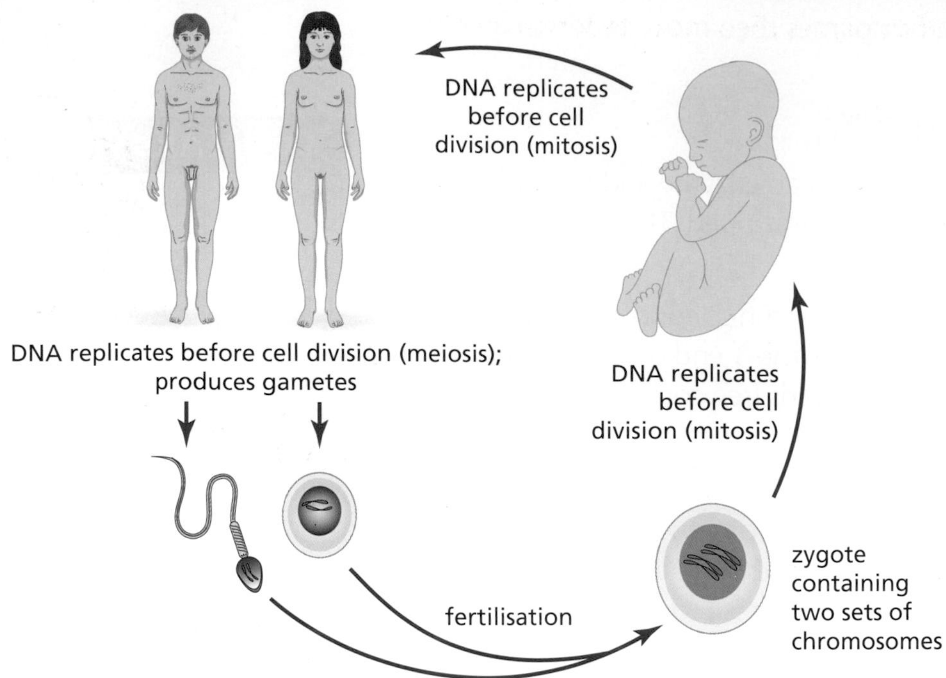

Figure 1.10 Importance of DNA replication in the human life cycle

Polymerase chain reaction

The polymerase chain reaction (PCR) is a laboratory technique that is used to amplify specific target sequences of DNA. The technique is carried out *in vitro*, which means that it happens outside the body of the organism in laboratory apparatus, and involves cycles of heating and cooling, as shown in Figure 1.11.

PCR involves exposing DNA to a series of temperature changes, known as thermal cycling. First the DNA is denatured between 92 °C and 98 °C, which separates the strands. Cooling to between 50 °C and 65 °C then allows complementary primers to bind to specific target sequences. The temperature is then raised to between 70 °C and 80 °C, when heat-tolerant DNA polymerase is used to synthesise new strands from free DNA nucleotides. These steps can be automated in a thermal cycling machine. Many repeated thermal cycles allow billions of copies of the target sequence to be produced.

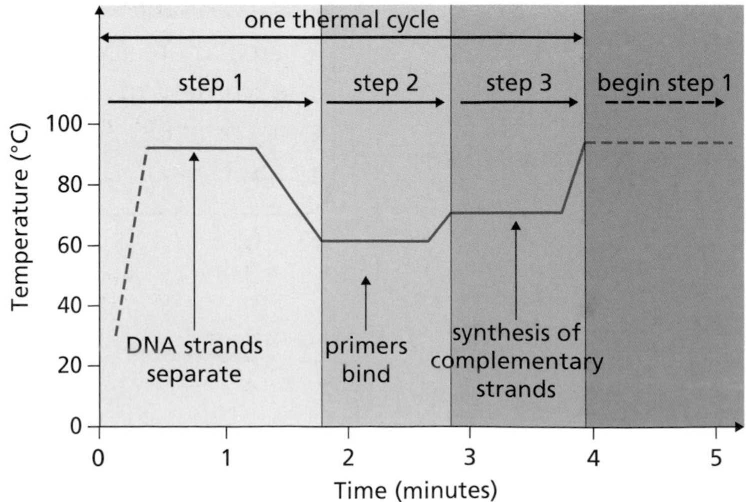

Figure 1.11 A summary of the events in the polymerase chain reaction

Hints & tips

Figure 1.10 illustrates the idea of vertical inheritance of genetic information from parents to offspring.

Key links

There is more about inheritance patterns in Key Area 1.7 (page 26).

Hints & tips

In vitro *means outside the body of an organism – the opposite of* in vivo, *which means inside the body. DNA replication is* in vitro *but PCR is* in vitro.

Hints & tips

In PCR, the primers locate a specific sequence of DNA, which is a bit like looking for a needle in a haystack. Once the DNA 'needle' has been found it can be amplified to make its own 'haystack' of copies.

Hints & tips

You should be able to calculate the number of DNA molecules present after a number of cycles in a PCR machine. Remember that after one cycle there are two molecules and the number doubles after every further cycle.

Figure 1.12 shows the pattern of amplification of a target sequence of DNA by the PCR process.

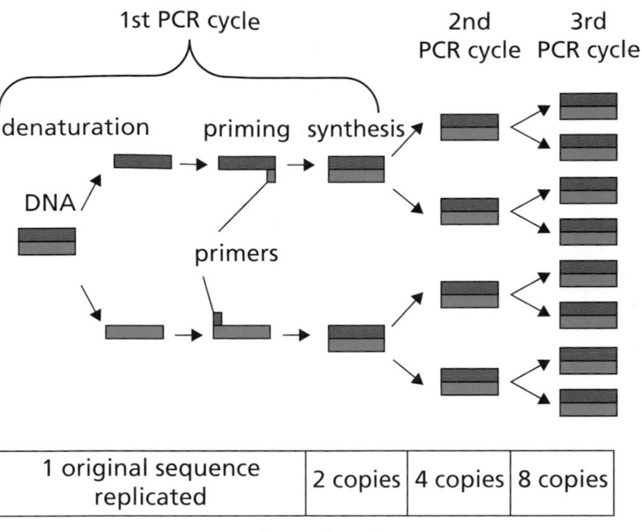

Figure 1.12 Amplification of a target sequence of DNA by PCR

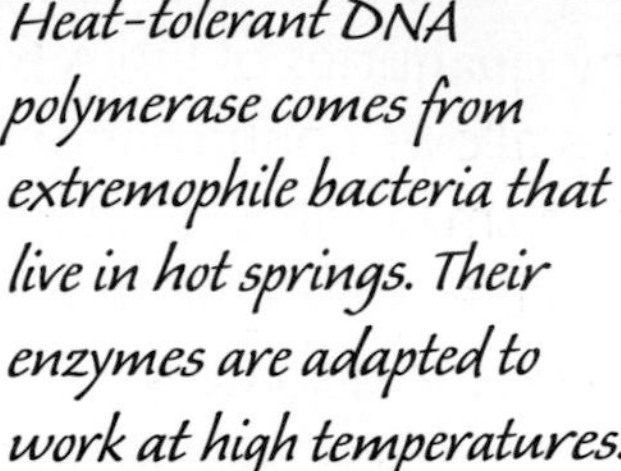

Key links

There is an example of a question on gel electrophoresis in the Skills of scientific inquiry chapter on page 146.

Gel electrophoresis

Gel electrophoresis is a technique which you need to become familiar with for your exam.

Gel electrophoresis is a method for separation of macromolecules, such as fragments of DNA, based on their size and electrical charge. Fragments are stained then placed in a well cut into a block of agarose gel. An electrical current is passed through the gel causing the fragments to move in the gel.

Shorter fragments move faster and further than longer ones. The final positions of the fragments are indicated by bands of stain in the gel, known as a ladder or profile as shown in Figure 1.13.

Key links

There is more about the action of endonucleases in Key Area 2.7 (page 84).

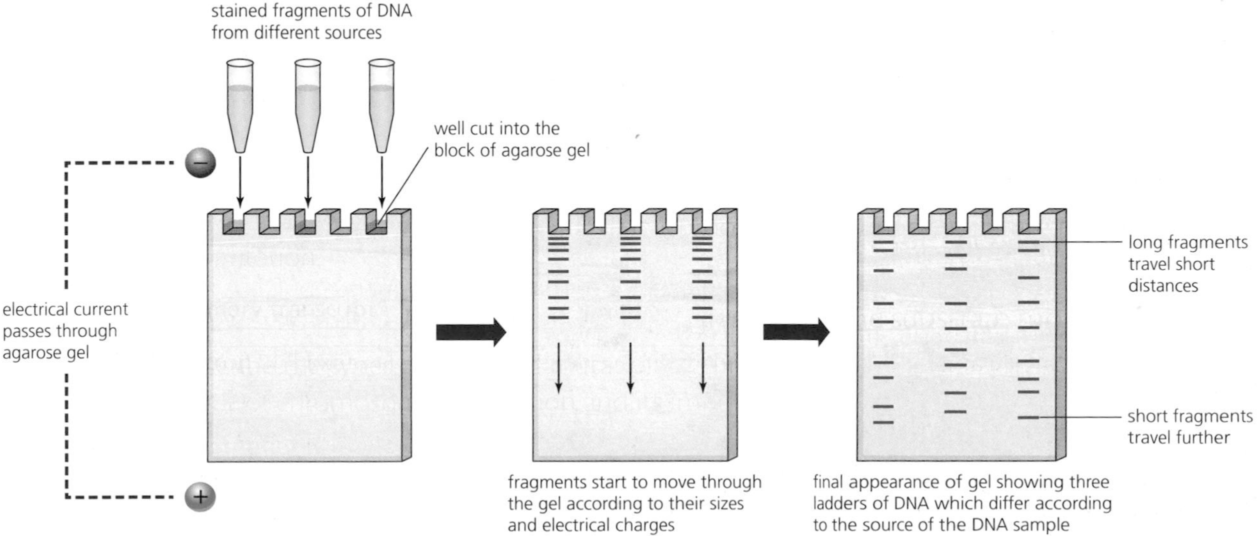

Figure 1.13 Diagram to show steps in gel electrophoresis

Applications of PCR

PCR has a variety of applications. In forensics it can be used to amplify tiny quantities of DNA from biological sources such as blood or semen. This allows confirmation of the presence of individuals at crime scenes from small samples of biological material. Other examples of the use of amplified DNA include settling paternity disputes and in the diagnosis of genetic disorders.

Hints & tips

Tiny samples of DNA from a crime scene can be amplified by PCR. The amplified samples are treated with endonucleases, stained, then passed through gel electrophoresis to produce unique DNA ladders called profiles or DNA fingerprints.

Key words

DNA polymerase – enzyme that adds free complementary DNA nucleotides during replication of DNA
Gel electrophoresis – technique used to separate macromolecules, such as DNA fragments of different sizes
Heat-tolerant DNA polymerase – enzyme from hot-spring bacteria, used in PCR
Lagging strand – DNA strand that is replicated in fragments
Leading strand – DNA strand that is replicated continuously
Ligase – enzyme that joins DNA fragments to make the lagging strand
Polymerase chain reaction (PCR) – method of amplifying sequences of DNA *in vitro*
Primer – short complementary strand of DNA
Replication – formation of identical copies of DNA molecules
Template strand – DNA strand on which a complementary copy is made

Questions

Short answer (1 or 2 marks)

1 The flow chart below shows temperature changes during steps in the polymerase chain reaction (PCR) procedure.

step 1 DNA heated to 95 °C	→	step 2 DNA cooled to 55 °C	→	step 3 DNA heated to 70 °C

a) Describe the effect of the increase in temperature at step 1 on the structure of DNA. (1)
b) State the need for the reduction in temperature at step 2. (1)
c) Explain why the DNA polymerase used in step 3 can function at 70 °C, although the high temperature would denature most enzymes. (1)
d) State the number of copies that would be present after one target sequence of DNA has passed through seven thermal cycles of PCR. (1)
e) Describe **one** application of the PCR procedure. (1)

Longer answer (3–10 marks)

2 Give an account of the replication of a molecule of DNA. (7)
T **3** Explain how PCR and gel electrophoresis can be used to identify the origin of a tiny sample of DNA from a crime scene. (5)

Answers are on page 36.

Key Area 1.3

Gene expression

Key points

1 **Gene expression** is the process by which specific **genes** are activated to produce a required protein. ☐
2 Gene expression involves the **transcription** and **translation** of DNA sequences. ☐
3 Genes are expressed to produce proteins. ☐
4 Only a fraction of the genes in a cell are expressed. ☐
5 Transcription and translation involves three types of **RNA** (mRNA, tRNA and rRNA). ☐
6 RNA is single stranded and is composed of nucleotides containing **ribose sugar**, phosphate and one of four bases: cytosine, guanine, adenine and **uracil**. ☐
7 **Mature messenger RNA (mRNA)** carries a copy of the DNA code from the nucleus to the **ribosome**. ☐
8 mRNA is transcribed from DNA in the nucleus and translated into proteins by ribosomes in the cytoplasm. ☐
9 Each triplet of bases on the mRNA molecule is called a **codon** and codes for a specific amino acid. ☐
10 **Transfer RNA (tRNA)** folds due to complementary base pairing. ☐
11 Each tRNA molecule carries its specific **amino acid** to the ribosome. ☐
12 **Ribosomal RNA (rRNA)** and proteins form the ribosome. ☐
13 A tRNA molecule has an **anticodon** (an exposed triplet of bases) at one end and an **attachment site** for a specific amino acid at the other end. ☐
14 DNA in the nucleus is transcribed to produce messenger RNA (mRNA), which carries a copy of the genetic code. ☐
15 In transcription, the enzyme **RNA polymerase** moves along DNA unwinding the double helix and breaking the hydrogen bonds between the bases. ☐
16 RNA polymerase synthesises a **primary transcript** of mRNA from RNA nucleotides by complementary base pairing. ☐
17 Uracil in RNA is complementary to adenine. ☐
18 Eukaryotic genes have **introns** (non-coding regions) and **exons** (coding regions). ☐
19 The introns of the primary transcript are non-coding regions and are removed. ☐
20 The exons are coding regions and are joined together to form the **mature transcript**. ☐
21 The order of the exons is unchanged during splicing. ☐
22 **RNA splicing** forms a mature mRNA transcript. ☐
23 The mature mRNA transcript carries a copy of the DNA code from the nucleus to the ribosomes, where it is translated. ☐
24 tRNA is involved in the translation of mRNA into a **polypeptide** at a ribosome. ☐
25 Most codons code for specific amino acids but there are also **start codons** and **stop codons**. Translation begins at a start codon and ends at a stop codon. ☐
26 Amino acids are carried by specific tRNA molecules. ☐ ⇨

27 tRNA anticodons align and bond to mRNA codons by complementary base pairing, translating the genetic code into a sequence of amino acids. ☐
28 **Peptide bonds** join the amino acids together to form a polypeptide. ☐
29 Following polypeptide formation, tRNA exits the ribosome to collect further amino acids. ☐
30 Different proteins can be expressed from one gene, as a result of **alternative RNA splicing**. ☐
31 Different mature mRNA transcripts are produced from the same primary transcript depending on which exons are retained. ☐
32 Amino acids are linked by peptide bonds to form polypeptides. ☐
33 Polypeptide chains fold to form the three-dimensional shape of a protein, held together by hydrogen bonds and other **molecular interactions** between individual amino acids. ☐
34 Proteins have a large variety of shapes which determines their functions. ☐
35 Gene expression results in proteins which determine the **phenotype** of an individual organism. ☐
36 **Environmental factors** also influence phenotype. ☐

Summary notes

The genetic code

The base sequence of DNA forms the genetic code. This code is found in all forms of life, which suggests that all life originated from a common ancestor.

Key links

There is more about the idea of common ancestors in Key Area 1.8 (page 32).

Genes are the units of genetic code that make up the genotype of an organism. These are expressed to produce proteins, which form the structure and control the functions of the organism. The phenotype of an individual organism is determined by the proteins produced by the expression of its genes. Only a fraction of the genes in a cell are expressed depending on the proteins required by that cell.

Figure 1.14 summarises how gene expression and environmental factors influence the phenotype of an organism.

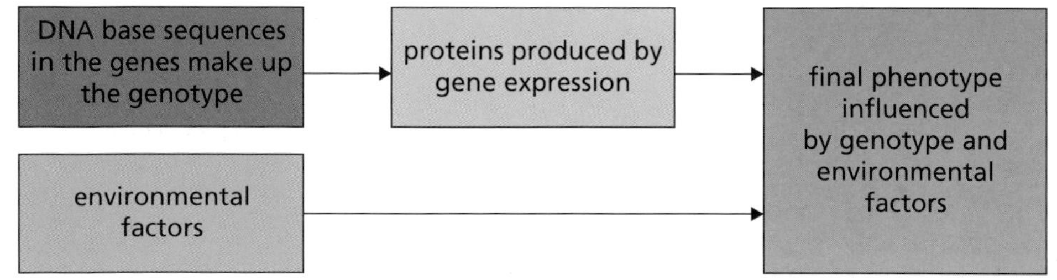

Figure 1.14 Flow chart to show how genotype and environmental factors influence final phenotype

Key links

There is more about which genes in a cell are expressed in Key Area 1.4 (page 17).

Structure and function of protein

Protein molecules are polypeptide chains (Figure 1.15). A polypeptide is a chain of amino acids held together by peptide bonds. The polypeptide is folded to give a protein with a three-dimensional shape held in place by hydrogen bonds and other interactions between individual amino acids. The shape of a protein is linked to its function.

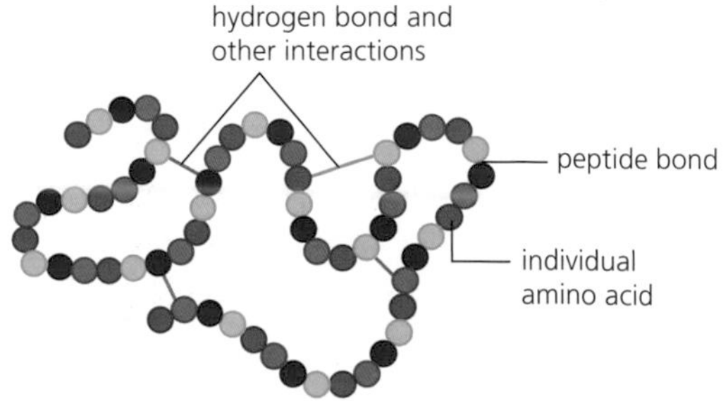

Figure 1.15 Diagram of a polypeptide chain folded to give a specific three-dimensional shape

Functions include structural components of cells and enzymes as shown in the following table.

Protein group	Function	Example
Structural components of cells	Building blocks of cell structure	Protein components of membranes such as the carrier proteins in the electron transport chain
Enzymes	Speed up the rate of chemical reactions	Enzymes to know about include DNA and RNA polymerase, ligase, endonucleases and RuBisCo

Key links

There is more about the electron transport chain in Key Area 2.2 (page 59).

Key links

There is more about the action of enzymes in Key Area 2.1b (page 51).

Stages of gene expression

Genes are expressed in two main stages, transcription and translation. In transcription, a copy of the gene in the form of a molecule called mRNA is created. In translation, a specific sequence of amino acids is built up using the mRNA code. These stages are shown in Figure 1.16.

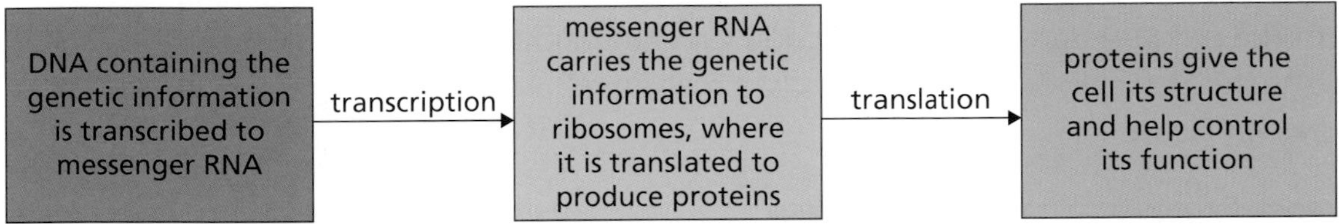

Figure 1.16 Relationships between the substances involved in gene expression

Ribonucleic acid (RNA)

Gene expression relies on various forms of RNA. RNA is very similar to DNA but has differences mainly in the nucleotides that it is made up of. DNA nucleotides have deoxyribose sugar but RNA nucleotides have ribose. DNA nucleotide bases are adenine, thymine, guanine and cytosine. In RNA, the base uracil (U) replaces thymine. Uracil also pairs with adenine in complementary base pairing. RNA is single stranded although there can be some base pairing of nucleotides. RNA nucleotides are shown in Figure 1.17.

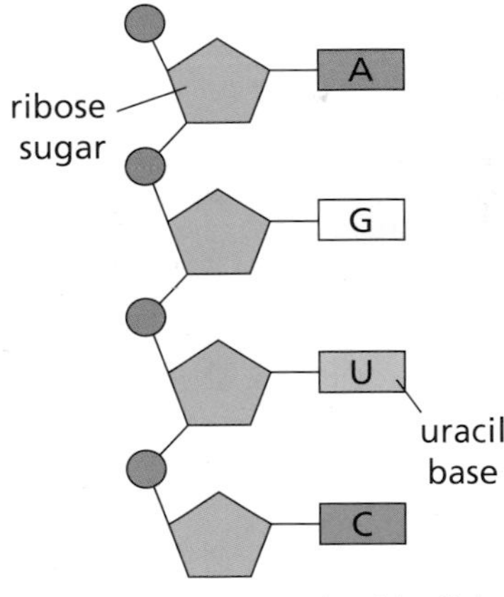

Figure 1.17 RNA nucleotides linked into a single strand, and showing the differences from DNA nucleotides

There are three types of RNA. Messenger RNA (mRNA) carries a complementary copy of the genetic code from the DNA in the nucleus to the ribosomes in the cytoplasm. Transfer RNA (tRNA) carries specific amino acids to ribosomes, where they can be assembled to form polypeptide chains. Ribosomal RNA (rRNA) is combined with proteins to make up the structure of ribosomes.

Transcription

In the first step, the enzyme RNA polymerase unwinds and unzips the double helix of the gene to be expressed and aligns free RNA nucleotides against the exposed DNA nucleotides of the template strand. Complementary base pairing ensures correct positioning of RNA nucleotides, which are then joined to form a primary transcript. The primary transcript is a complementary copy of the gene made up of groups of three bases called codons, as shown in Figure 1.18. Primary transcripts start with start codons and finish with stop codons.

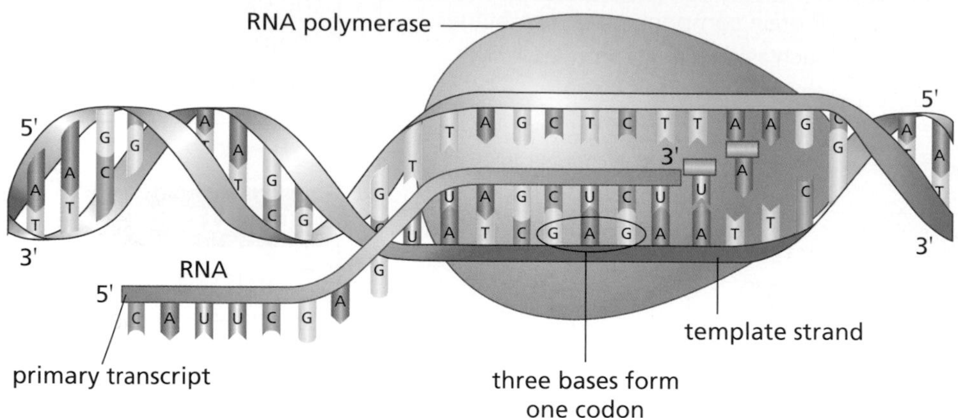

Figure 1.18 Formation of a primary transcript

Each transcript has introns and exons. Introns are non-coding regions and are removed from the primary transcript, leaving coding regions known as exons. The exons are then spliced together to form a mature mRNA transcript, as shown in Figure 1.19.

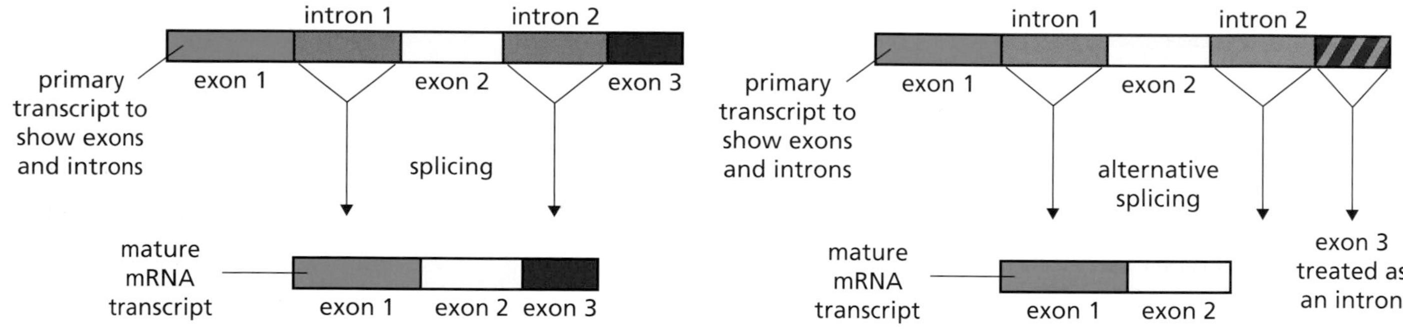

Figure 1.19 RNA splicing

The alternative splicing shown in Figure 1.19 allows a primary transcript to form different mRNA molecules depending on which exons are included in the mature mRNA. Exons are always kept in the same order after splicing. In this way, a DNA sequence may be expressed in different ways in cells of different tissues or cells of different ages.

Ribosomes and translation

In eukaryotes mRNA molecules move from the nucleus to the ribosomes to be translated.

Ribosomes made of rRNA and protein are found free in the cytoplasm of cells or bound to the membranes of its endoplasmic reticulum (ER). A mature mRNA molecule binds onto a ribosome. mRNA carries a start codon, which starts translation, and a stop codon, which causes translation to finish when the polypeptide is complete.

The folded tRNA molecules, held by complementary base pairs, have a triplet of three bases called an anticodon and an attachment site to transport a specific amino acid to mRNA on the ribosomes, as shown in

Figure 1.20. They are recognised and align with the mRNA according to their anticodons, which are complementary to the codons of mRNA. The amino acids that have been lined up bind through peptide bonds to form polypeptides. The polypeptide folds to form a protein, which is held together by hydrogen bonds and other molecular interactions between amino acids.

Translation is summarised in Figure 1.20.

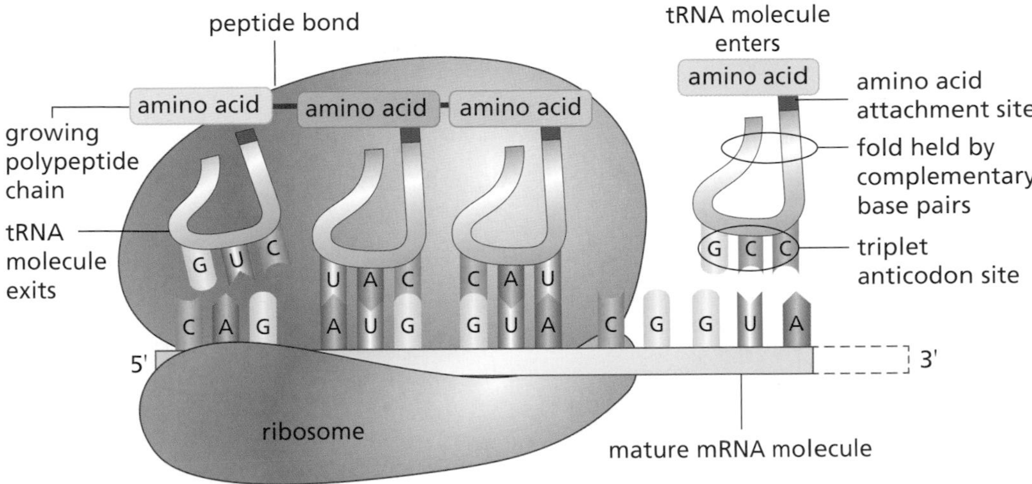

Figure 1.20 Translation of mRNA to form a polypeptide chain

One gene, many proteins

Different proteins can be expressed from the same gene due to alternative RNA splicing. This idea is summarised in Figure 1.21.

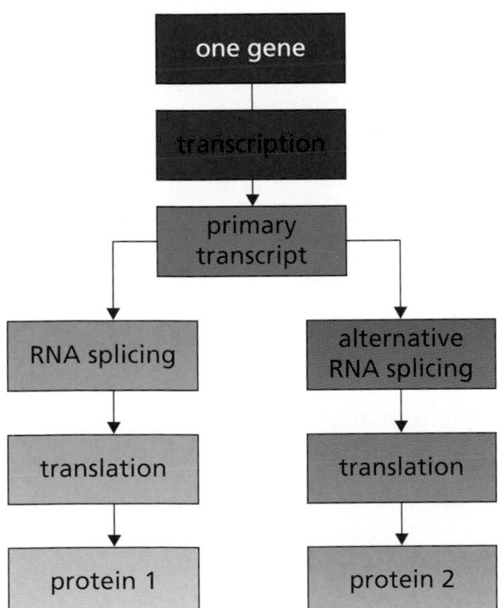

Figure 1.21 Production of different proteins from one gene

Hints & tips

It is important to remember that the sequences of codons on mature mRNA lead to the production of a specific polypeptide.

Key words

Alternative RNA splicing – synthesis of different mature transcripts from the same primary transcript
Amino acid – unit of polypeptide structure
Anticodon – sequence of three bases on tRNA that specifies an amino acid
Attachment site – position on a tRNA molecule at which a specific amino acid binds
Codon – sequence of three bases on mRNA that specifies an amino acid
Environmental factors – these include light, temperature, nutrients and other factors which can affect the phenotype of an organism
Exon – sequence of DNA that codes for part of a protein
Gene – DNA sequence which codes for a protein
Gene expression – transcription and translation
Intron – non-coding sequence of DNA within a gene
Mature messenger RNA (mRNA) – carries a copy of the DNA code to a ribosome
Mature transcript – alternative term for mature mRNA
Molecular interactions – various chemical links joining amino acids and giving protein molecules their shape
Peptide bonds – strong chemical links which join amino acids in the primary structure of polypeptides
Phenotype – outward appearance of an organism
Polypeptide – short strand of amino acids
Primary transcript – molecule made when DNA is transcribed
Ribosomal RNA (rRNA) – type of RNA that makes up ribosomes
Ribose sugar – sugar component of an RNA nucleotide
Ribosome – site of protein synthesis; composed of rRNA and protein
RNA – ribonucleic acid, which occurs in several forms in cells
RNA polymerase – enzyme involved in synthesis of primary transcripts from DNA
RNA splicing – joining of exons following the removal of introns from a primary transcript
Start codon – triplet transcribed from DNA to a primary transcript indicating the start of the gene
Stop codon – triplet on the primary transcript which signals a stop to translation
Transcription – copying of a DNA sequence to make a primary transcript
Transfer RNA (tRNA) – transfers specific amino acids to the mRNA on the ribosomes
Translation – production of a polypeptide using sequences of mRNA
Uracil – RNA base not found in DNA but complementary to adenine in transcription

Questions ?

Short answer (1 or 2 marks)

1 Proteins are chains of amino acids folded into three-dimensional shapes.
  a) Name the bonds that hold the amino acids together in sequence. (1)
  b) Describe how the chains of amino acids are held in their three-dimensional shape. (1)
  c) Explain the importance of the three-dimensional shape of a protein molecule. (1)
  d) Give **two** examples of the groups of proteins found in living cells. (2)
2 Give **one** way in which the expression of a single gene can result in different proteins being produced. (1)
3 Eukaryotic genes are made up from base sequences known as introns and exons.
  State how introns and exons differ. (1)
4 Describe the functions of start and stop codons. (2)

Longer answer (3–10 marks)

5 Give an account of gene expression in eukaryotic cells under the following headings:
  a) transcription of DNA (4)
  b) translation of mature mRNA. (4)

Answers are on page 37.

Key Area 1.4

Cellular differentiation

Key points !

1 **Cellular differentiation** is the process by which a cell expresses certain genes to produce proteins that are characteristic for that type of cell. This allows a cell to carry out specialised functions. ☐
2 **Meristems** are regions of unspecialised cells in plants that can divide (self-renew) and/or differentiate. ☐
3 **Stem cells** are unspecialised cells in animals that can divide (self-renew) and/or differentiate. ☐
4 Cells in the very early embryo can differentiate into all the cell types that make up the organism and so are **pluripotent**. ☐
5 All the genes in **embryonic stem cells** can be switched on so these cells can differentiate into all the cell types that make up the organism. ☐
6 **Tissue stem cells** are involved in the growth, repair and renewal of the cells found in that tissue. They are **multipotent**. ☐
7 Tissue stem cells are multipotent as they can differentiate into all of the types of cell found in a particular tissue type. For example, blood stem cells located in bone marrow can give rise to all types of blood cell. ☐
8 Stem cells have both **therapeutic** and **research** uses. ☐
9 Therapeutic uses involve the repair of damaged or diseased organs or tissues. They are used in corneal repair and the regeneration of damaged skin. ☐
10 Stem cells from the embryo can self-renew, under the right conditions, in the lab. ☐
11 In research, stem cells are used as model cells to study how diseases develop and are employed in drug testing. ☐
12 Stem cell research provides information on how cell processes, such as cell growth, differentiation and gene regulation, work. ☐
13 Use of embryonic stem cells can offer effective treatments for disease and injury; however, it raises ethical issues because it involves the destruction of an early embryo. ☐

Summary notes

Cellular differentiation

Differentiation is a process by which unspecialised cells become specialised for a specific function in the body of a living organism. The process depends on the control of gene expression. Specialised cells express the genes characteristic of that cell type. A mammalian muscle cell expresses mammalian muscle cell genes and so on. Therefore mammalian muscle cells produce mammalian muscle cell proteins. This idea is shown in Figure 1.22.

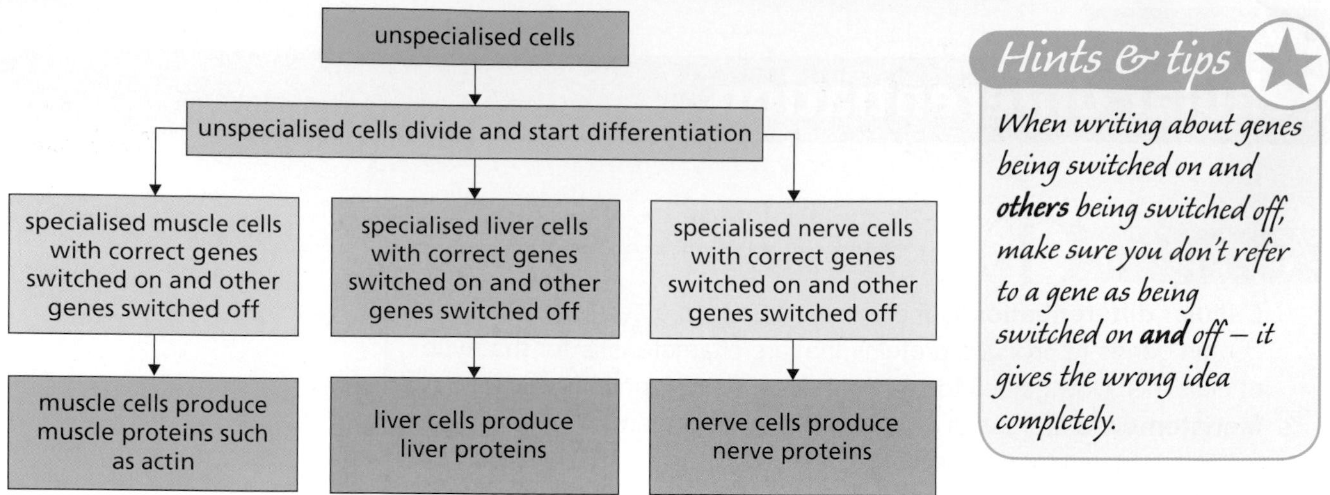

Figure 1.22 Differentiation

Hints & tips

When writing about genes being switched on and ***others*** *being switched off, make sure you don't refer to a gene as being switched on* ***and*** *off – it gives the wrong idea completely.*

Meristems

Regions in plants containing unspecialised cells are called meristems. Meristem cells are capable of cell division and produce new cells that can then differentiate to produce permanent tissues. The meristems of a flowering plant are shown in Figure 1.23.

shoot apical meristems (in buds)
lateral meristems
root apical meristems

Figure 1.23 Plant seedling, showing the positions of meristems

Stem cells

Stem cells are relatively unspecialised cells in animals. They can divide to produce cells that can then differentiate into various cell types and more stem cells. In early embryos, embryonic stem cells differentiate into all the cell types that make up the adult organism, as shown in Figure 1.24. They are said to be pluripotent.

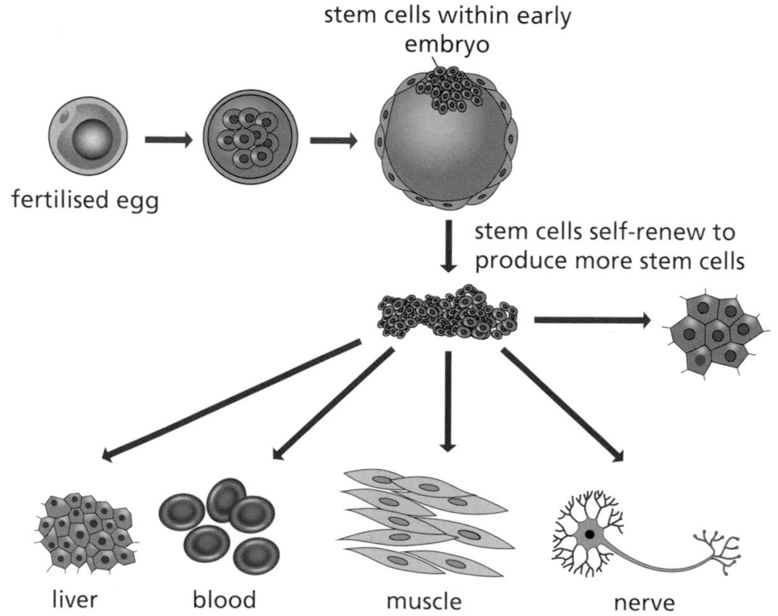

Various differentiated cells can be produced following the division of stem cells.

Figure 1.24 Pluripotent embryonic stem cells

In adults, stem cells within tissues differentiate to replace damaged cells of the type in those tissues, as shown in Figure 1.25. They are said to be multipotent.

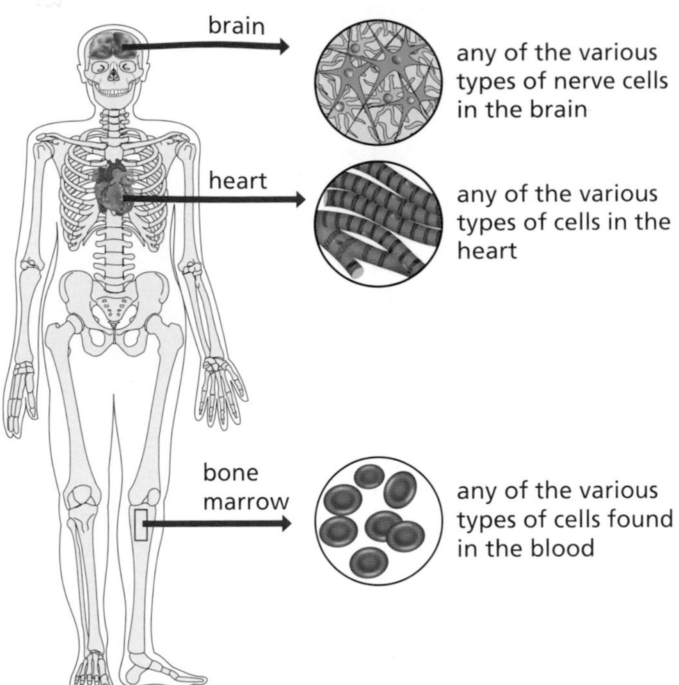

Figure 1.25 Multipotent tissue stem cells

Hints & tips

A key property of both stem cells and meristem cells is that they can divide. This allows them to produce a supply of identical cells, some of which can then differentiate.

Stem cell research and the therapeutic uses of stems

Stem cells can be used therapeutically to repair damaged or diseased organs and tissue. Stem cell research provides information on how cell processes such as growth, differentiation and gene regulation work. Stem cells can also be used as model cells to study how diseases develop or for drug testing. There are various ethical issues raised by stem cell research. The following table summarises some of these issues.

Ethical question	Notes
Is the prevention of suffering more important than the duty to preserve human life?	Embryonic stem cell research gives us a moral dilemma. It forces us to choose between two moral principles important to humans: our duty to prevent or ease suffering and our duty to respect the value of human life. Which is more important?
Is there a possibility of stem cells being used eugenically?	Embryonic stem cells might be used to change the body characteristics of already healthy and well individuals. Should this be allowed?
Could stem cells become part of an illegal trade in biological material?	It is possible that stem cells might be bought and sold illegally and treatment could become subject to the ability of an individual to pay. Should this be allowed?

Hints & tips

You must know the difference between therapeutic and research uses of stem cells. Therapeutic uses involve the actual treatment of patients.

Embryonic stem cell research could lead to new medical treatments, which could save human lives and relieve human suffering. On the other hand, to obtain embryonic stem cells, an early-stage embryo has to be destroyed, meaning the loss of a potential human life. Which moral principle should be followed in this situation? Does the answer lie in our attitude to the embryo? Does it have the status of a person? What do you think about this issue?

Hints & tips

Ethical issues are difficult. Make sure that you are aware of why the issues exist – you could be asked to give an example of an issue in your exam.

Key words

Cellular differentiation – changes to cells involving switching on certain genes and switching off others
Embryonic stem cell – stem cell from an embryo that can divide and become any type of cell
Meristem – region in a plant in which mitosis occurs
Multipotent – refers to stem cells which can differentiate into all of the types of cell found in a particular tissue
Pluripotent – refers to embryonic stem cells which can differentiate into all the cell types that make up the organism
Research use – use as a model for study or for testing of drugs to develop new treatments
Stem cell – cell that can divide and then differentiate in animals
Therapeutic use – used as part of medical therapy
Tissue (adult) stem cell – stem cell from tissue that can divide and differentiate to become cells of that tissue

Questions

Short answer (1 or 2 marks)

1 Describe how an embryonic stem cell differs from a tissue stem cell. (2)
2 Describe the importance of tissue stem cells in the human body. (2)

Longer answer (3–10 marks)

3 Give an account of the process of cellular differentiation in animals and the function of meristems in plants. (4)
4 Give an account of ethical issues related to stem cell use. (4)
5 Outline the differences between the research and therapeutic uses of stem cells. (4)

Answers are on page 37.

Key Areas 1.5 and 1.6

Structure of the genome and mutation

Key points

1. The **genome** of an organism is its entire hereditary information encoded in DNA. ☐
2. An organism's genome is made up of genes, which are DNA sequences that code for protein, and other DNA sequences that do not code for proteins. ☐
3. Most of the genome in eukaryotes consists of **non-coding sequences**. ☐
4. Non-coding sequences include those that regulate transcription and those that are transcribed into RNA but are not translated. ☐
5. tRNA and rRNA are non-translated forms of RNA. ☐
6. **Mutations** are changes in the DNA that can result in no protein or an altered protein being synthesised. ☐
7. **Single gene mutations** involve the alteration of a DNA nucleotide sequence as a result of the **substitution**, **insertion** or **deletion of nucleotides**. ☐
8. Nucleotide substitutions include **missense**, **nonsense** and **splice-site mutations**. ☐
9. Missense mutations result in one amino acid being changed for another. This may give a non-functional protein or have little effect on the protein. ☐
10. Nonsense mutations result in a premature stop codon being produced, which results in a shorter protein. ☐
11. Splice-site mutations result in some introns being retained and/or some exons not being included in the mature transcript. ☐
12. Nucleotide insertions or deletions result in **frame-shift mutations**. ☐
13. Frame-shift mutations cause all of the codons and all of the amino acids after the mutation to be changed. This has a major effect on the structure of the protein produced. ☐
14. Chromosome structure mutations that involve alterations to the structure of a chromosome include **duplication**, **deletion**, **inversion** and **translocation**. ☐
15. Duplication allows potential beneficial mutations to occur in a duplicated gene, while the original gene can still be expressed to produce its protein. ☐

Summary notes

Structure of the genome

The genome of an organism is the total genetic information encoded into the base sequence of its DNA. The genome contains those sequences that code for protein (genes) and those that do not. The non-coding sequences include introns, those that regulate transcription, and those that are transcribed into RNA but not translated. The table on page 22 summarises genome structure.

Part of genome	Function of sequences
Coding sequences (genes)	Code for amino acid sequences in proteins
Non-coding sequences	Regulate transcription by turning genes on or off
	Transcribed but not translated (e.g. rRNA, tRNA)
	Introns within genes

Organisms pass copies of their genome to their offspring.

Mutation

Mutations are rare, random changes to DNA sequences. The following table shows different types of mutation and their effects.

Mutation	Effect
… of single genes	No protein or an altered protein is produced
… at splice sites	Can cause introns to be left in mature mRNA, leading to an altered protein
… of chromosomes	Affects structure of chromosomes present in cells

Hints & tips

*Remember **ROLF** – mutations are of random occurrence and low frequency.*

Single gene mutations

Single gene mutations occur within genes and involve alteration of a DNA nucleotide sequence. Gene mutations result in no protein or an altered protein being expressed, as shown in the following table.

Hints & tips

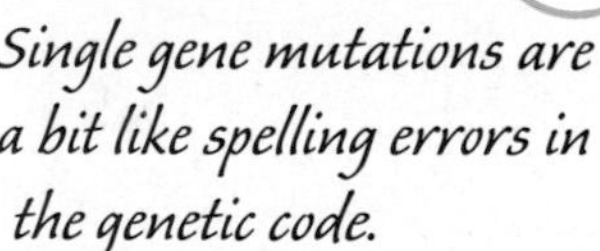

Single gene mutations are a bit like spelling errors in the genetic code.

Hints & tips

*Remember **SID** –*
***S**ubstitution*
***I**nsertion*
***D**eletion*

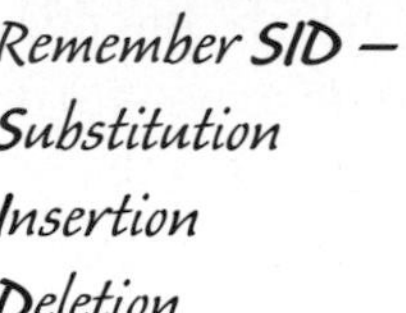

Single gene mutation	Description	Example of nucleotide base sequence changes	Impact on protein structure
Substitution	A single nucleotide removed from a DNA sequence and replaced by another with a different base	Normal sequence: … ATGTCCATG … Following mutation: … ATG**G**CCATG …	Minor impact, since there is a maximum of one amino acid changed in the protein structure – called **missense** Major impact could result if mutation results in production of a premature stop codon – called **nonsense**
Insertion	Additional nucleotide added into a DNA sequence	Normal sequence: … ATGTCCATG … Following mutation: … ATGT**G**CCATG …	Major effect on protein likely since all amino acids coded for after the mutation could be affected Sometimes called **frame-shift** mutations
Deletion	Nucleotide removed from a DNA sequence but not replaced with another	Normal sequence: … ATGTCCATG … Following mutation: … ATGTC_ATG …	

Splice-site mutation

A single gene mutation at a splice site could result in an intron being left in the mature mRNA or an exon being left out. This could result in an altered protein that would not function normally, as shown in Figure 1.26.

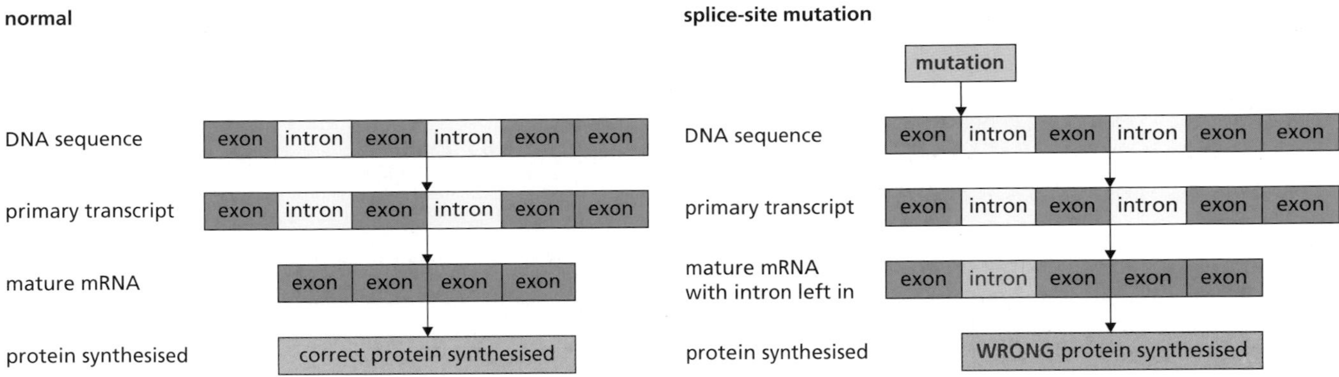

Figure 1.26 Effect of a splice-site mutation on a protein in which an intron is left in the mature mRNA

Significance of gene mutation in evolution

Mutation provides the only source of new variation for living organisms and so has had vital importance in evolution. An example involving a single gene mutation of the gene for human haemoglobin is outlined in the following example box.

Example

Sickle cell disease

Haemoglobin is a protein that carries oxygen in the blood. Each person has two copies of the haemoglobin gene, each of which codes for half of their haemoglobin.

About 5000 years ago, a single gene mutation of the gene is thought to have arisen in Africa. This involved a substitution, which resulted in a form of haemoglobin with one amino acid different from the normal form. This abnormal haemoglobin leads to the collapsing of the red blood cells that contain it when conditions become acidic during exercise. The collapsed cells stick in narrow blood vessels, causing sickle cell disease.

Individuals with two copies of the mutation have a seriously debilitating condition called sickle cell anaemia. Individuals with one copy of the normal gene and one copy of the mutated gene have sickle cell trait and show less severe symptoms of the disease. In some parts of Africa, it is an advantage to have sickle cell trait because it protects individuals from malaria parasites. The parasites live inside red blood cells and their acidic waste products cause cells to collapse, which targets them for destruction by white blood cells. The mutation is now common in parts of Africa and demonstrates the importance of mutation in evolution.

Key links

There is more about selection pressure in Key Area 1.7 (page 26) and parasites in Key Area 3.5 (page 124).

Chromosome mutations

Some mutations affect the structure of chromosomes present in the cells of living organisms, as shown in the table below. They arise when pieces of one chromosome break off and are lost or join back into the chromosome complement in a different way.

Chromosome mutation	Description	Diagram	Example or effect
Deletion	Detached genes are lost completely	ABCDE FGH —deletion→ ABCE FGH	Cri du chat syndrome in humans involves a loss of part of chromosome 5
Inversion	Chromosome breaks in two places and a set of genes rotates through 180°	ABCDE FGH (180°) —inversion→ ADCBE FGH	One cause of haemophilia A is an inversion within a blood-clotting gene
Translocation	Detached genes become attached to a **non-homologous chromosome** in the complement	ABCDE FGH, MNOPQ R —translocation→ MNOCDE FGH, ABPQ R	One type of Down syndrome in humans is caused in this way
Duplication	A set of genes from one chromosome becomes attached to its **homologous partner**, leading to repeated genes	ABCDE FGH —duplication→ ABCBCDE FGH	Some duplications can be highly detrimental; others can be important in evolution

Chromosome mutations and evolution

It is thought that duplicated genes can undergo single gene mutations without affecting the functioning of the original copy of the gene. A new potentially beneficial gene could then appear, giving an organism a selective advantage without any effect on the functioning of the original gene.

One example is found in some coldwater fish. They have an antifreeze protein in their blood that has allowed colonisation of extremely cold water. The gene that encodes this protein seems to have been formed by the mutation of a gene coding for a vital digestive enzyme but, because of duplication, the modern fish still expresses the digestive enzyme as well as having the antifreeze.

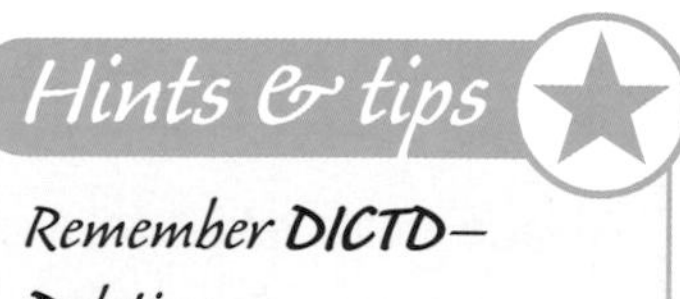

Remember ***DICTD***–
D*eletion*
I*nversion*
C*hromosome mutation*
T*ranslocation*
D*uplication*

Key words

Deletion – chromosome mutation in which a sequence of genes is lost from a chromosome
Deletion of nucleotides – single gene mutation involving removal of a nucleotide from a sequence
Duplication – chromosome mutation in which a section of chromosome is added from its homologous partner
Frame-shift mutation – single gene mutations which cause all codons and all amino acids after the mutation to be changed
Genome – total genetic material present in an organism
Insertion – single gene mutation in which an additional nucleotide is placed into a sequence
Inversion – chromosome mutation in which a set of genes rotates through 180°
Missense – refers to a single gene mutation which results in one amino acid in a protein being changed for another
Mutations – random changes to DNA sequences
Non-coding sequence – DNA sequence that does not encode protein
Nonsense – refers to a single gene mutation which results in premature stop codon being produced
Single gene mutations – mutations which involve the alteration of a sequence of DNA nucleotide
Splice-site mutation – mutation at a point where coding and non-coding regions meet in a section of DNA
Substitution – single gene mutation in which one nucleotide is replaced by another
Translocation – mutation in which part of a chromosome becomes attached to a non-homologous chromosome

Questions ?

Short answer (1 or 2 marks)

1 The table below shows the effects of some single gene mutations on base sequences in DNA.

Original base sequence of gene	Number	Effect of mutation on base sequence
... TTACGCTAC ...	1	... TACGCTAC ...
	2	... TTACGGCTAC ...
	3	... TGACGCTAC ...

a) Name mutations 1–3. (2)
b) Describe the effects of mutations 1 and 3 on the structure of the polypeptide coded for by the original gene. (2)

2 Describe the importance of single gene mutation in evolution. (2)

3 Give the meaning of the following types of mutation:
a) nonsense mutation (1)
b) splice-site mutation. (1)

Longer answer (3–10 marks)

4 Name and describe the types of structural mutation of chromosomes. (4)

5 Give an account of chromosome duplication and its importance in evolution. (3)

Answers are on page 38.

Key Area 1.7

Evolution

Key points

1. **Evolution** is the result of changes in organisms over generations due to genomic variations. ☐
2. **Natural selection** is the non-random increase in frequency of DNA sequences that increase survival and the non-random reduction in the frequency of **deleterious sequences**. ☐
3. Changes in phenotype frequency can arise as a result of **stabilising**, **directional** and **disruptive selection**. ☐
4. In stabilising selection, an average phenotype is selected for and extremes of the phenotype range are selected against. ☐
5. In directional selection, one extreme of the phenotype range is selected for. ☐
6. In disruptive selection, two or more phenotypes are selected for. ☐
7. Natural selection is more rapid in prokaryotes. ☐
8. Prokaryotes can exchange genetic material horizontally, resulting in faster evolutionary change than in organisms that only use vertical transfer. ☐
9. **Horizontal gene transfer** is where genes are transferred between individuals in the same generation. ☐
10. **Vertical gene transfer** is where genes are transferred from parent to offspring as a result of sexual or asexual reproduction. ☐
11. A **species** is a group of organisms capable of interbreeding and producing fertile offspring, and which does not normally breed with other groups. ☐
12. **Speciation** is the generation of new **biological species** by evolution as a result of **isolation**, mutation and selection. ☐
13. Isolation barriers are important in preventing gene flow between sub-populations during speciation. ☐
14. **Geographical barriers** prevent gene flow and lead to **allopatric speciation.** ☐
15. **Behavioural barriers** and **ecological barriers** prevent gene flow and lead to **sympatric speciation.** ☐

Summary notes

Gene transfer

Genomic sequences can be inherited by vertical or horizontal gene transfer.

Vertical gene transfer

This means that the genetic material is passed from parent to offspring either sexually or asexually when a species reproduces, as shown in Figure 1.27 (a) and (b).

Horizontal gene transfer

Prokaryotes can inherit genomic sequences horizontally. This means that genes are transferred between individuals in the same generation, as shown in Figure 1.27 (c). This method of transfer can result in faster evolutionary change than in organisms that use only vertical transfer.

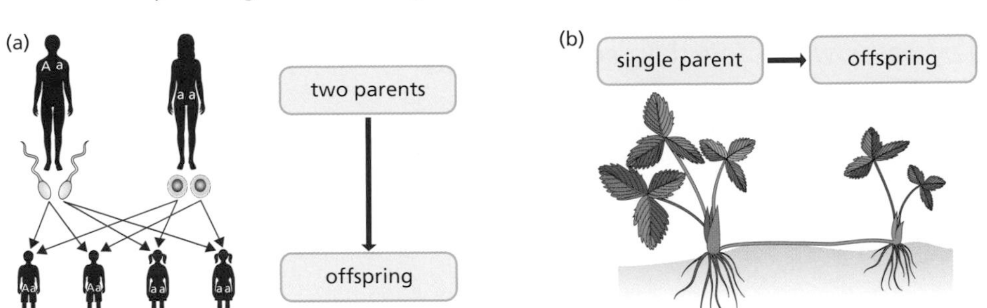

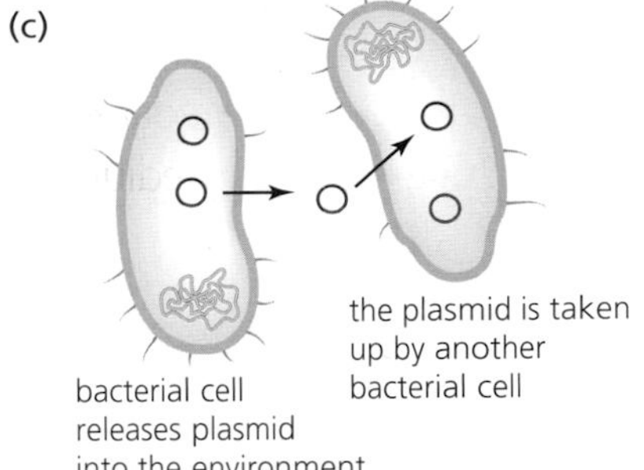

Figure 1.27 Vertical inheritance **(a)** Sexual – genetic sequences passed from male and female parents to offspring **(b)** Asexual – genetic sequences passed from single parent to offspring; Horizontal inheritance **(c)** An example of horizontal gene transfer in prokaryotes

Hints & tips

Don't be tempted to identify vertical or horizontal inheritance patterns according to how diagrams are laid out – a horizontal arrow in a diagram does not necessarily mean horizontal inheritance!

Evolution by natural selection

Changes in genomic sequences in organisms result in changes to the species over time – this is called evolution. The changes depend on random alterations in the genome due to mutation. Mutations can be advantageous or detrimental to the survival of the organism.

Advantageous mutations help organisms to survive to reproduce, and so the mutation is passed on to offspring. Natural selection is the process which leads to non-random increase in the frequency of DNA sequences that increase survival and the non-random reduction in unfavourable (deleterious) sequences (Figure 1.28).

Evolution can be more rapid in prokaryotes, which can exchange genetic material horizontally, compared to species that transfer genetic material only vertically.

For example, genes for antibiotic resistance that have evolved in one species of bacteria can be transferred to another species of bacteria horizontally, so increasing their rate of evolutionary change.

Key links 

There is coverage of the artificial horizontal transfer of genomic sequences in prokaryotes in Key Area 2.7 (page 84).

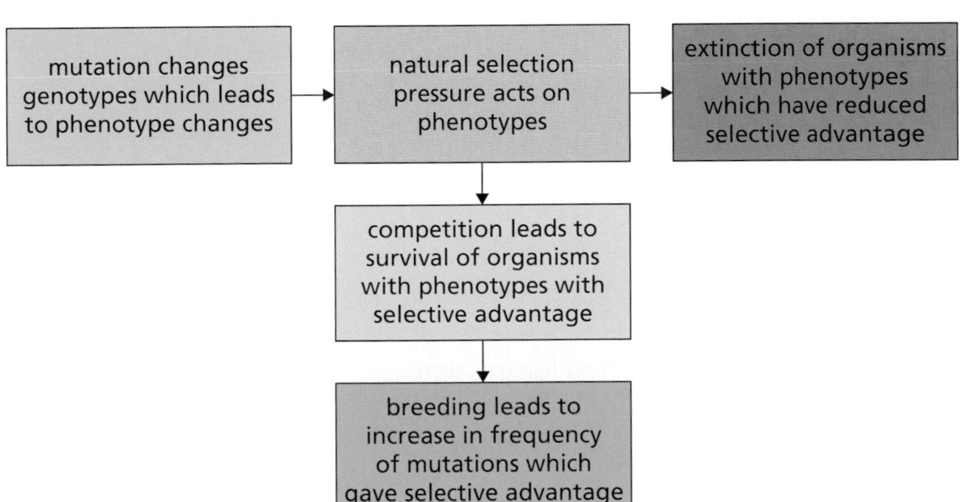

Figure 1.28 Summary of links between mutations and survival in evolution

Type of selection pressure

Stabilising selection

In stabilising selection, average phenotypes are selected for and extremes of the phenotype are selected against. An example of this is found in gall flies, which produce galls on plants that contain their offspring. Flies that produce small galls often have these parasitised by wasps, while those that produce large galls often have them predated by woodpeckers. There is selection pressure to produce galls of a medium size, as shown in the graph in Figure 1.29.

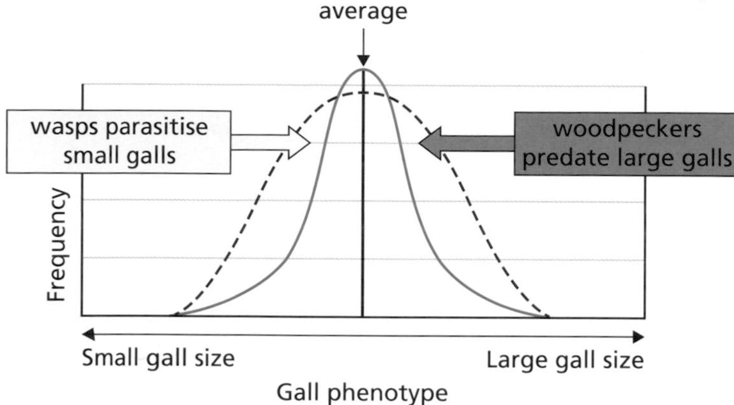

Figure 1.29 Stabilising selection pressure on gall fly gall sizes

Directional selection

In directional selection, one extreme of the phenotype is selected for. In cliff swallows, for example, larger body size is a selective advantage, as shown in the chart in Figure 1.30.

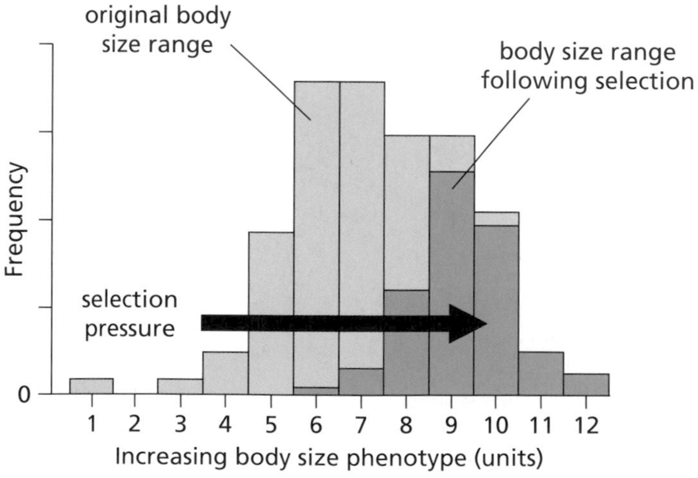

Figure 1.30 Directional selection for larger body size in cliff swallows

Disruptive selection

In disruptive selection, two or more phenotypes are selected for. In salmon, for example, larger male fish are better able to compete for territories, but smaller male fish without territories are able to sneak into those of larger fish and fertilise eggs without being detected, as shown in Figure 1.31.

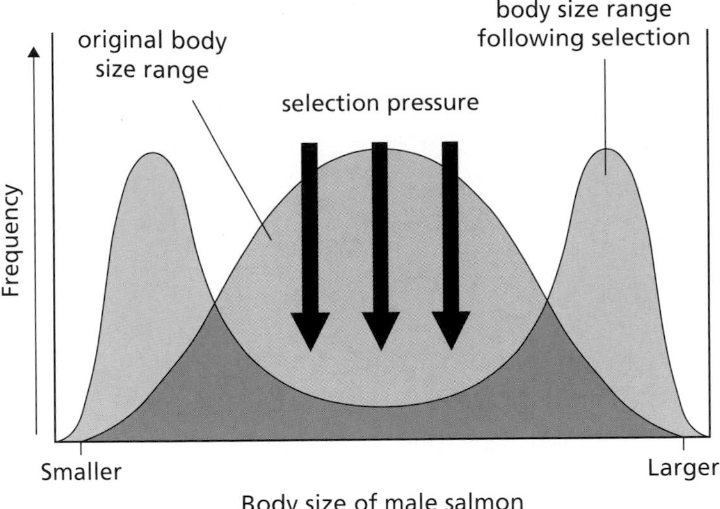

Figure 1.31 Disruptive selection in salmon

Speciation

A species is a group of very similar interbreeding organisms that give rise to fertile offspring. The biological species is the basic unit of classification. Its members are reproductively isolated so do not normally attempt to interbreed with other species. Speciation is the evolution of two or more species from a common ancestor.

New biological species are produced by natural selection, usually acting over long periods of time in an evolutionary process called speciation.

Allopatric speciation

In allopatric speciation, sub-populations of a species become isolated from each other by geographical barriers such as mountains or oceans. No gene flow occurs between the sub-populations and they build up separate genetic differences based on natural selection acting on different mutations in their different environments.

Hints & tips

Remember that different species do not produce fertile young when they interbreed. They can only produce fertile young with members of the same species.

Example

Figure 1.32 shows an example of allopatric speciation involving Galapagos finches.

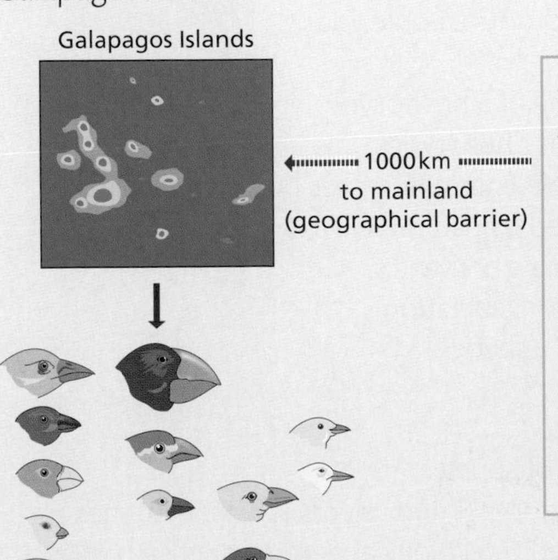

Mainland Ecuador

Ancestral finches from mainland Ecuador migrated to the Galapagos Islands and became isolated from each other. Mutations arose differently within different isolated populations and natural selection favoured different beak shapes according to the food supply and habitats available. After millions of years, evolution caused the radiation into the many different species seen today.

Figure 1.32 Allopatric speciation in Galapagos finch species

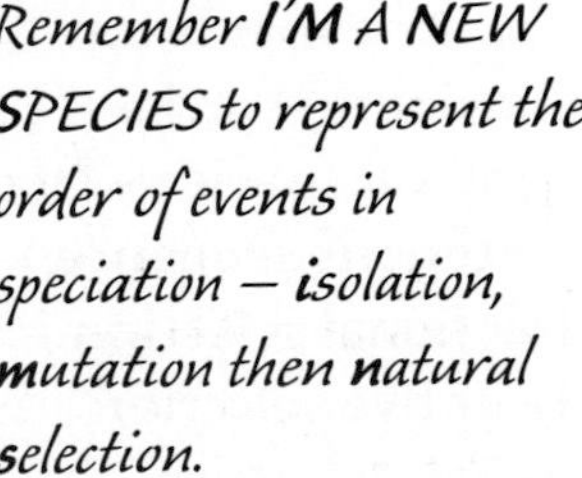
Hints & tips

*Remember **I'M A NEW SPECIES** to represent the order of events in speciation – **isolation**, **mutation** then **natural selection**.*

Sympatric speciation

In sympatric speciation, isolation occurs when gene flow between populations is stopped by ecological or behavioural barriers to the exchange of genetic sequences. The sub-populations live side by side but do not interbreed and so natural selection is able to act separately on them.

Example

American maggot flies

The ancestors of American maggot flies laid their eggs only in native hawthorn berries. The species now lays eggs in hawthorn berries and in apples, which were introduced to America about 200 years ago.

Females generally choose to lay their eggs in the type of fruit in which they grew up. Males tend to look for mates on the type of fruit in which they grew up. So hawthorn flies tend to mate with other hawthorn flies and apple flies tend to mate with other apple flies, as shown in Figure 1.33. This means that gene flow between parts of the population that mate on different types of fruit is reduced. This may be the first step towards sympatric speciation.

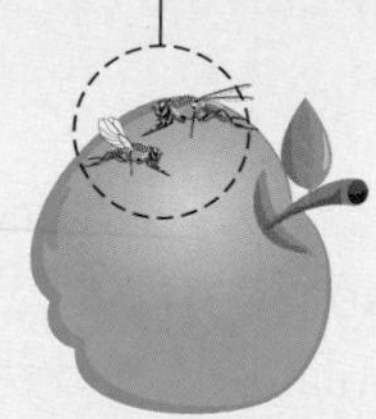

Figure 1.33 Sympatric isolation in American maggot flies

Key words

Allopatric speciation – speciation in which gene flow is prevented by a geographical barrier
Behavioural barrier – barrier to gene flow caused by behavioural differences between individuals
Biological species – group of similar organisms interbreeding to produce fertile young
Deleterious sequence – DNA sequence that lowers survival rate
Directional selection – natural selection that favours one extreme of a phenotype
Disruptive selection – natural selection that favours two different phenotypes
Ecological barrier – barrier to gene flow caused by ecological preference differences between individuals
Evolution – changes to organisms over time that are mainly caused by natural selection
Geographical barrier – physical barrier to gene flow, such as a mountain or river
Horizontal gene transfer – inheritance of genetic material within a generation
Isolation – refers to prevention of gene flow between populations of a species
Natural selection – process that ensures survival of the fittest
Speciation – evolutionary process by which new species are formed
Species – group of organisms which interbreed to produce fertile offspring
Stabilising selection – natural selection that favours average phenotypes and selects against extremes
Sympatric speciation – speciation in which gene flow is prevented by ecological or reproductive barriers
Vertical gene transfer – inheritance of genetic material from parents by offspring

Questions ?

Short answer (1 or 2 marks)

1 Gene sequences are inherited vertically from parent to offspring.
Describe how genetic sequences are inherited horizontally. (2)

2 Describe the differences between the isolation barriers in allopatric and sympatric speciation. (2)

3 Explain what is meant by the term *species*. (1)

Longer answer (3–10 marks)

4 Give an account of the different types of natural selection pressure involved in evolution. (6)

5 Give an account of the role of natural selection in evolution. (5)

6 Give an account of allopatric speciation. (5)

Answers are on page 38.

Key Area 1.8

Genomic sequencing

Key points

1. In **genomic sequencing** the sequence of nucleotide bases can be determined for individual genes and entire genomes. ☐
2. Computer programs can be used to identify base sequences by looking for sequences similar to known genes. ☐
3. To compare sequence data, computer and statistical analyses (**bioinformatics**) are required. ☐
4. Comparison of genomes from different species reveals that many genes are highly conserved across different organisms. ☐
5. Many genomes have been sequenced, particularly of disease-causing organisms, pest species and species that are important **model organisms** for research. ☐
6. **Phylogenetics** is the study of evolutionary history and relationships. ☐
7. Evidence from phylogenetics and **molecular clocks** has been used to determine the main sequence of events in evolution. ☐
8. The sequence of events can be determined using **sequence data** and **fossil evidence**. ☐
9. Sequence data is used to study the evolutionary relatedness among groups of organisms. ☐
10. Sequence divergence is used to estimate time since lineages diverged. ☐
11. Comparison of sequences provides evidence of the three **domains of life – Bacteria, Archaea** and Eukarya (eukaryotes). ☐
12. Sequence data and fossil evidence have been used to determine the main sequence of events in evolution of life: cells, **last universal ancestor**, prokaryotes, photosynthetic organisms, eukaryotes, multicellularity, animals, vertebrates, land plants. ☐
13. Molecular clocks are used to show when species diverged during evolution. ☐
14. Molecular clocks assume a constant mutation rate and show differences in DNA sequences or amino acid sequences. Therefore differences in sequence data between species indicate the time of divergence from a **common ancestor**. ☐
15. An individual's genome can be analysed to predict the likelihood of developing certain diseases. ☐
16. **Pharmacogenetics** is the use of genome information in the choice of drugs. ☐
17. An individual's personal genome sequence can be used to select the most effective drugs and dosage to treat their disease (**personalised medicine**). ☐

Summary notes

Genomics and phylogenetics

Sequences of nucleotide bases can be determined for individual genes and entire genomes. Amino acid sequences can be determined for individual proteins. The results are called sequence data. To compare sequence data, computer and statistical analyses are required. These techniques are known as bioinformatics.

Sequence data can be used to study evolutionary relatedness among groups of organisms. Sequence divergence can be used along with the fossil record in drawing molecular clock graphs, such as that for cytochrome C or haemoglobin. Molecular clock diagrams are based on the assumption that the mutation rate of genes leading to amino acid differences in proteins is constant through time.

Figure 1.34 shows how differences in amino acid sequence in haemoglobin can confirm the evolutionary relatedness of vertebrate groups.

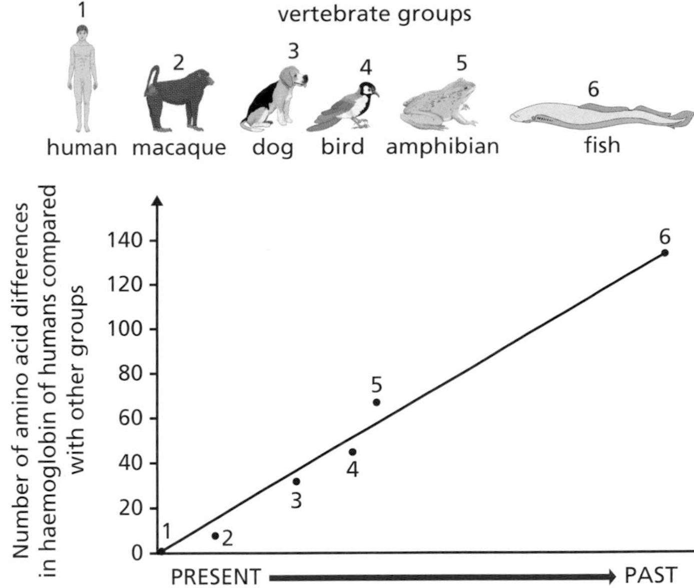

Figure 1.34 Molecular clock for haemoglobin relative to human haemoglobin

Phylogenetic trees

Sequence data results can be used in combination with fossil evidence to draw phylogenetic trees, as shown in Figure 1.35. In this figure, letter x shows the last common ancestor of A and B and letter y shows the last common ancestor of C, D and E.

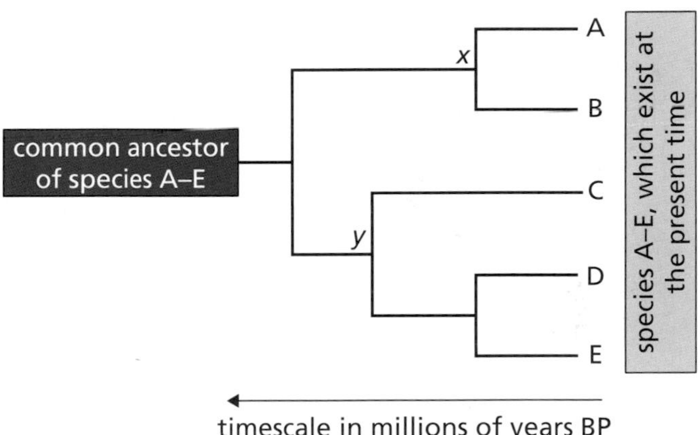

Figure 1.35 The basic features of a phylogenetic tree

The tree in Figure 1.36 shows the emergence of the three fundamental domains of life on Earth today – Bacteria, Archaea and Eukarya (eukaryotes).

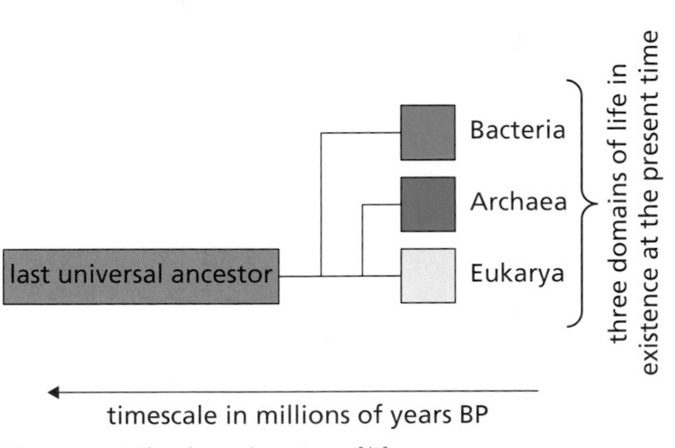

Figure 1.36 The three domains of life

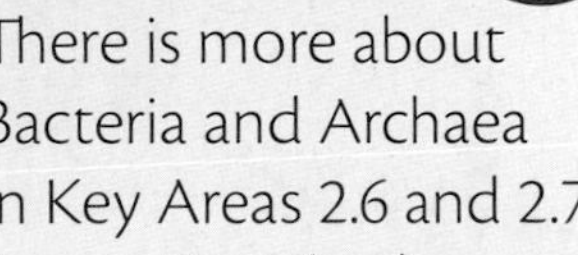

There is more about Bacteria and Archaea in Key Areas 2.6 and 2.7 (page 78 and 84).

Sequence data combined with fossil evidence is used to determine the main sequence of events in the evolution of life, as shown in the table below. Comparison of genomes from different species show that many genes are found in a conserved state across different organisms.

Event in evolution	Approximate time (million years BP)	Notes
appearance of prokaryotic cells	3600	first bacteria-like cells with no distinct nucleus
existence of the last universal ancestor (LUA)	3500	existence of the most recent organism from which all organisms now living on Earth have descended
photosynthesis	3400	first organisms which could use light energy in the synthesis of complex molecules
appearance of eukaryotic cells	2000	first organisms with cells containing a true nucleus
appearance of multicellular organisms	1000	first organisms whose bodies were composed of a group of interdependent cells
appearance of animals	600	first multicellular eukaryotic organisms which ingested other organisms as food
appearance of vertebrate animals	540	first animals which possessed backbones
appearance of land plants	475	first plants to live in terrestrial (land) habitats

Personal genomics and health

It is possible to sequence the genome of an individual human being. Many diseases have a genetic risk component and so analysis of an individual's genome could lead to personalised medicine through increased information on the likelihood of a treatment being successful in a specific individual. Pharmacogenetics refers to the study of how drugs might be designed to best suit individuals with particular genetic sequences.

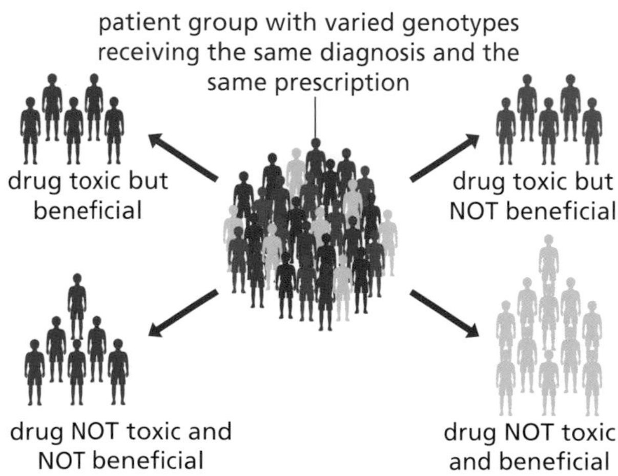

Figure 1.37 The effects of the genotype of a group of patients with liver disease

Figure 1.37 shows the effects of the genotype of a group of patients with liver disease on treatment with a new, potentially beneficial drug – this highlights the problem. If patient genotype could be related to the effects of the drug before treatment, this could maximise the benefits of the treatment by identification of the group for whom the drug would be most beneficial.

Key words

Archaea – a domain of life containing single-celled microorganisms
Bacteria – a domain of life
Bioinformatics – use of computers and statistics in analysis of sequence data
Common ancestor – species from which more than one more modern species have diverged
Domains of life – three major groups, Bacteria, Archaea and Eukarya, into which all known living species can be divided
Fossil evidence – information derived from the remains of extinct organisms
Genomic sequencing – procedure used to reveal the nucleotide sequence of a genome
Last universal ancestor – the most recent organism from which all other organisms have descended
Model organisms – species which have had their genomes sequenced and can be used as references
Molecular clock – graph that shows differences in sequence data for a protein against time
Personalised medicine – possible future development in which treatment is based on an individual's genome
Pharmacogenetics – the study of inherited differences affecting metabolic pathways which can affect individual therapeutic responses to, and side-effect of, drugs
Phylogenetics – study of evolutionary relatedness of species
Sequence data – information concerning amino acid or nucleotide base sequences

Questions ?

Short answer (1 or 2 marks)

1 Give the **two** types of information needed to construct phylogenetic trees. (2)
2 Give the term applied to the statistical and mathematical treatment of sequence data. (1)
3 Name the **three** domains of life. (2)

Longer answer (3–10 marks)

4 Give an account of personalised genomics and medicine. (3)
5 Give an account of construction and limitations of molecular clocks. (5)
6 Give an account of the main events in the evolution of life on Earth. (5)

Answers are on page 39.

Answers

Key Area 1.1

Short answer questions

1 a) double helix [1]
   b) by hydrogen bonding; between complementary bases [1 each = 2]
   c) a deoxyribose sugar molecule; a phosphate group; a base [all 3 = 2; 2/1 = 1]
   d) (i) adenine pairs with thymine; guanine pairs with cytosine [1 each = 2]
      (ii) each strand runs in the opposite direction to its complementary strand; reference to 3′ and 5′ [1 each = 2]

2 Their cells contain plasmids; eukaryotes don't normally contain plasmids/prokaryotes normally contain plasmids [1 each = 2]

3 In prokaryotic bacteria DNA is packaged in circular chromosomes; and in plasmids; in eukaryotic plant cells DNA is packaged in linear chromosomes in a nucleus; and in circular chromosomes in chloroplasts; and in circular chromosomes in mitochondria [any 4 = 4]

Longer answer questions

4 Function: carries inherited information; in a chemical language; in its base sequence [any 2 = 2]

Structure: double helix; chains/strands of nucleotides; nucleotide is deoxyribose, a phosphate and a base; sugar–phosphate backbone; complementary base pairing *or* A with T, G with C; antiparallel chains/strands [any 5 = 5]

[total = 7]

Answers

Key Area 1.2

Short answer questions

1 a) denatures/separates DNA strands [1]
   b) allows primers to bind [1]
   c) it is heat tolerant/comes from bacteria adapted to live in hot springs [1]
   d) 128 copies [1]
   e) amplifies DNA to help solve crimes/settle paternity disputes/diagnose genetic disorders (there are many other acceptable answers) [any 1 = 1]

Longer answer questions

2 DNA unwinds; hydrogen bonds between complementary bases are broken; primers bind at 3′ end of lead template strand; DNA polymerase adds complementary DNA nucleotides to lead strand continuously; primers bind to lagging strand in many places; DNA polymerase adds complementary DNA nucleotides to lagging strand in fragments; fragments joined by ligase; replication requires energy/ATP [any 7 = 7]

3 treat sample with PCR to amplify the DNA; treat amplified DNA with endonucleases; stain DNA fragments; add stained fragments to a gel; apply an electric current to gel; to produce a DNA profile [any 5 = 5]

Answers

Key Area 1.3

Short answer questions

1 a) peptide bonds [1]
b) by hydrogen bonds *and* other linkages [1]
c) allows the protein to carry out its function [1]
d) structural; enzymes [1 each = 2]

2 alternative RNA splicing [1]

3 introns are non-coding regions *and* exons are coding regions [1]

4 start codons cause RNA polymerase to start transcription/cause a ribosome to start synthesising a polypeptide chain; stop codons cause transcription to stop/bring polypeptide chain synthesis to a halt [1 each = 2]

Longer answer questions

5 a) DNA unwinds and unzips; RNA polymerase adds complementary RNA nucleotides; (to make a) primary transcript; introns removed; exons spliced to make mature mRNA/a mature transcript; occurs in nucleus [any 4 = 4]
b) mRNA goes to ribosome; tRNA carries specific amino acids; anticodons on tRNA aligned with codons on mRNA; amino acids aligned in correct sequence; amino acids linked by peptide bonds [any 4 = 4] [total = 8]

Answers

Key Area 1.4

Short answer questions

1 embryonic stem cells divide and then differentiate into any type of cell/are pluripotent; tissue stem cells also divide but then only differentiate into cells of the type found in the tissue from which they came/are multipotent [1 each = 2]

2 divide/self-renew and differentiate to replace cells; which were/had become damaged/diseased cells [1 each = 2]

Longer answer questions

3 in tissue cells some genes switched off and other genes switched on; producing proteins characteristic of that tissue; allow tissue to carry out its function [any 2 = 2]

meristems contain unspecialised cells; meristem cells divide/self-renew; to produce cells which can differentiate to produce mature plant tissues [any 2 = 2] [total = 4]

4 to obtain embryonic stem cells an embryo has to be sacrificed; but using embryonic stem cells can potentially provide treatment for disease/injury; stem cells could be used to treat healthy individuals; stem cells could be part of an illegal trade; other ethical issue explained [any 4 = 4]

5 therapeutic uses involve treatment/repair of damaged/diseased organs/tissues; examples cornea/skin graft/bone marrow transplant [1 each]; research uses involve study of cell process/growth/division/differentiation OR gene regulation; can involve drug testing; can be used as model cells to study how diseases develop [any 2, 1 each] [total = 4]

Answers

Key Areas 1.5 and 1.6

Short answer questions

1 a) 1 deletion; 2 insertion; 3 substitution [all 3 = 2; 2/1 = 1]

b) 1/deletion gives frame-shift effect with major changes to amino acid sequence *or* all amino acids changed after the mutation [1]

3/substitution results in missense and gives minor changes to polypeptide *or* only one amino acid changed *or* results in nonsense if stop codon affected [1]

2 single gene mutation provides the variation needed for evolution/natural selection; natural selection favours mutations that increase survival [1 each = 2]

3 a) mutations that cause a polypeptide to be too short/translation to stop prematurely [1]

b) mutation that can result in mRNA with the wrong introns or exons [1]

Longer answer questions

4 deletion; involves loss of section of chromosome; translocation; involves section of one chromosome joining to another; duplication; involves a section of one chromosome being copied from its homologous partner; inversion; involves a set of genes rotating through 180°
[any 2 names = 2 and matching effects = 2]
[total = 4]

5 in duplication a second copy of a section of a chromosome is present; a single gene mutation in a duplicated region of a chromosome can produce an advantageous gene; original gene still present so can still be expressed/still produce its protein [1 each = 3]

Answers

Key Area 1.7

Short answer questions

1 passed from one prokaryotic organism to another; genes transferred between individuals in the same generation [1 each = 2]

2 allopatric speciation involves geographical barriers; sympatric speciation involves behavioural/ecological barriers [1 each = 2]

3 a group of organisms which can interbreed to produce fertile young [1]

Longer answer questions

4 stabilising selection – individuals with average phenotypes favoured; directional selection – individuals with an extreme phenotype favoured; disruptive selection – individuals with two or more different phenotypes favoured
[1 for each named type of selection and 1 for each correct matching definition = 6]

5 natural selection is non-random; individuals of a species vary; there is competition between individuals of the species; natural selection favours the survival of best-adapted individuals; they survive to reproduce and pass their beneficial genes to their offspring; this increases the frequency of the beneficial genes; after long periods of time new species may form [any 5 = 5]

6 geographical isolation barrier splits a population into sub-populations; barrier prevents gene flow; mutations occur randomly on different sides of the barrier; selection pressures are different on different sides of the barriers; natural selection acts over a long period; new species are formed on different sides of the barrier [any 5 = 5]

Answers

Key Area 1.8

Short answer questions

1 DNA/protein sequences; fossil record [1 each = 2]

2 bioinformatics [1]

3 Bacteria; Archaea; eukaryotes [all 3 = 2; any 2 = 1]

Longer answer questions

4 personalised genomics involves sequencing the genome of an individual; genomic differences are important in the effectiveness of treatments/drugs; so different treatments/drugs/dosages can be designed for an individual [3]

5 molecular clocks are based on amino acid sequence difference in a protein; molecular clocks are based on base sequence differences in DNA; sequence differences are graphed on the y-axis; time of divergence is graphed on the x-axis; mutation rate assumed to be constant; mutation rates may vary through time [any 5 = 5]

6 appearance of cells, last universal ancestor, prokaryotes, photosynthesis, eukaryotes, multicellularity, animals, vertebrates, land plants [any 5]

Practice course assessment: Area 1 – DNA and the genome (50 marks)

Paper 1 (10 marks)

1 In the chloroplasts of plant cells, genetic material is organised into

A linear chromosomes **B** RNA molecules
C circular chromosomes **D** circular plasmids.

2 The graph below shows temperature changes involved in one cycle of the polymerase chain reaction (PCR).

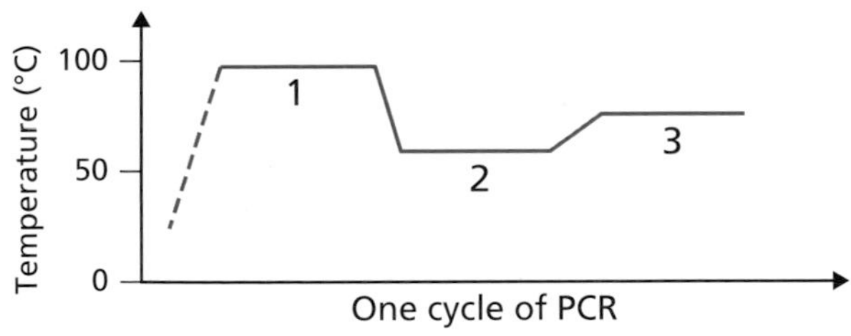

Which row in the table below identifies the main events occurring in each region of the graph?

	Main event occurring during PCR cycle		
	1	2	3
A	Primers bind to target sequences on DNA	DNA polymerase replicates regions of DNA	DNA strands separate
B	DNA strands separate	DNA polymerase replicates regions of DNA	Primers bind to target sequences on DNA
C	DNA strands separate	Primers bind to target sequences on DNA	DNA polymerase replicates regions of DNA
D	Primers bind to target sequences on DNA	DNA strands separate	DNA polymerase replicates regions of DNA

3 The list shows steps in the synthesis of the protein mucin.

1 Transcription of DNA
2 Formation of peptide bonds
3 RNA splicing
4 Translation of mRNA

In which order would these steps occur?

A 1 3 4 2
B 1 4 2 3
C 3 1 2 4
D 3 4 2 1

4 Which of the following substances are required in the replication of the lagging strand of a DNA molecule?

A DNA polymerase and ligase only
B DNA polymerase and primers only
A Ligase and primers only
D DNA polymerase, ligase and primers

Questions 5 and 6 refer to the graph on the right, which shows changes in the number of human stem cells present in a culture over a period of 16 days. Also shown is the level of activity of an enzyme found in stem cell cytoplasm over the same period.

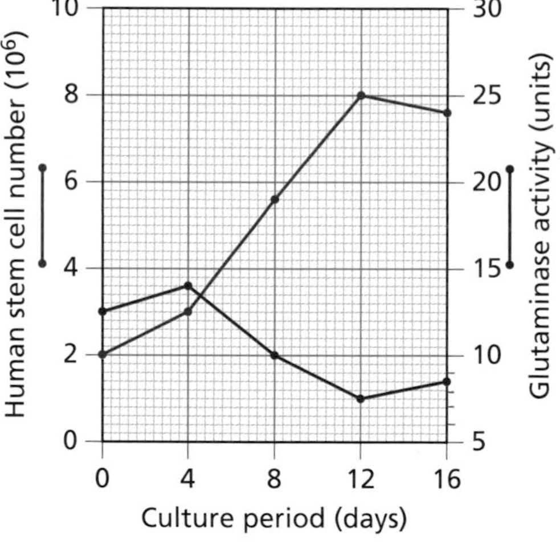

5 How many units of enzyme activity were recorded when the number of cells was at 25 per cent of its maximum over the 16-day period?

A 2
B 3
C 10
D 12.5

6 What was the percentage decrease in enzyme activity between days 8 and 12 of the period?

A 25
B 50
C 100
D 200

7 The following are events in the evolution of life on Earth.

1 Animals appear
2 Vertebrates appear
3 Land plants appear

In which order are these events thought to have occurred?

A 1 2 3
B 1 3 2
C 3 1 2
D 3 2 1

8 Bioinformatics is the

A production of phylogenetic tree diagrams
B development of personalised medicine
C use of mathematical and statistical techniques in genetic sequencing
D construction of molecular clocks in evolutionary studies.

9 The charts below show the effects of selection on the body mass of male birds in a population of barn swallows. Measurements were made before and after a period of extremely high temperatures during their breeding season.

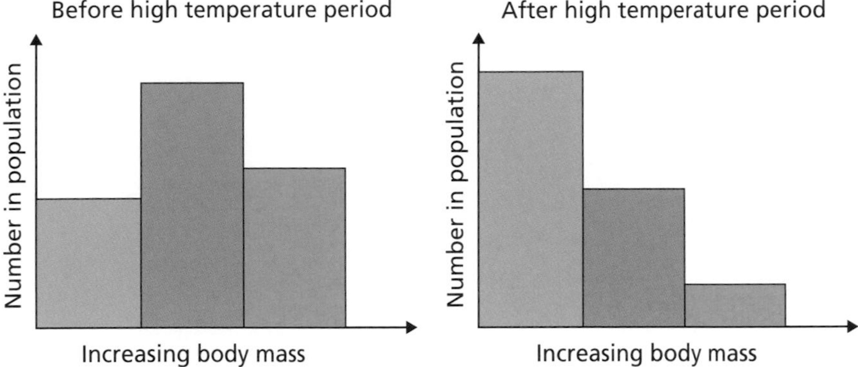

The occurrence of which process is indicated by the results?

A stabilising selection B allopatric speciation

C sympatric speciation D directional selection

10 Which of the following is a gene mutation that can have frame-shift effects?

A substitution B deletion

C duplication D translocation

Paper 2 (40 marks)

1 The diagram below shows a short section of a DNA molecule.

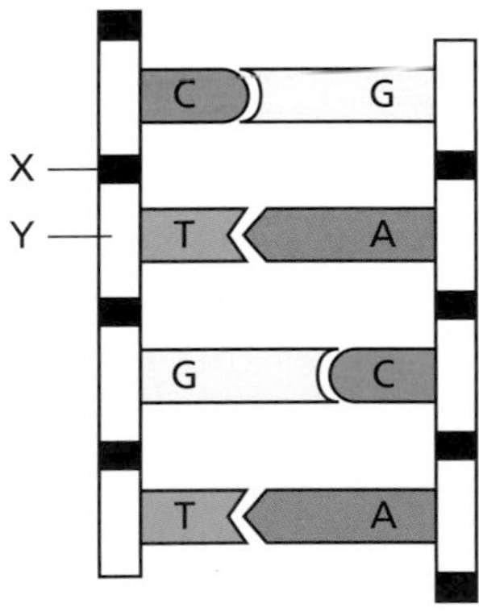

a) Name components X and Y, which make up the backbone of one of the strands of the molecule. (2)
b) The two strands are antiparallel. Describe what is meant by this term. (1)
c) Give the term which describes the shape of a DNA molecule that is **not** shown by this diagram. (1)
d) Name the type of bond that connects the bases in DNA. (1)

2 The diagram below shows a primary transcript that is processed to produce a mature messenger RNA transcript.

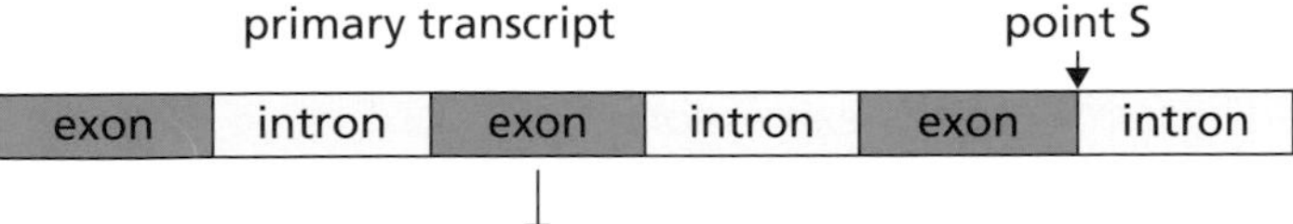

a) Name the enzyme that adds RNA nucleotides to form a primary transcript. (1)
b) Describe what happens to the primary transcript to produce the mature messenger RNA transcript. (2)
c) Explain the effect of a mutation at point S on the structure of the polypeptide synthesised. (1)
d) Name the sites in a cell where proteins are synthesised. (1)

3 The diagram below shows the sequence of bases in some mRNA codons and the amino acids for which they code.

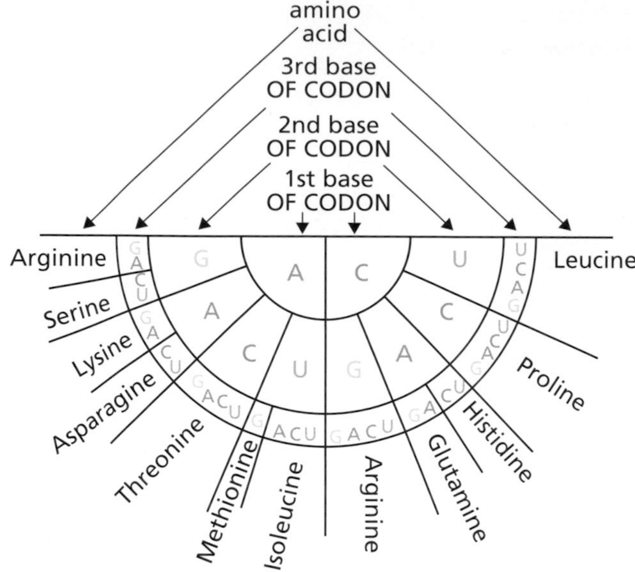

a) Give the mRNA codon that codes for the amino acid methionine. (1)

b) Name all the amino acids that are specified by codons that have cytosine (C) as their second base. (1)

c) Name the amino acid that could be carried by a tRNA molecule with the anticodon GUA. (1)

d) A mutation caused a codon to change from AGG to CGG. Explain the impact this would have on the structure of the protein produced as a result and explain your answer. (2)

4 The diagram below shows the location of a meristem and some mature tissues in a plant root.

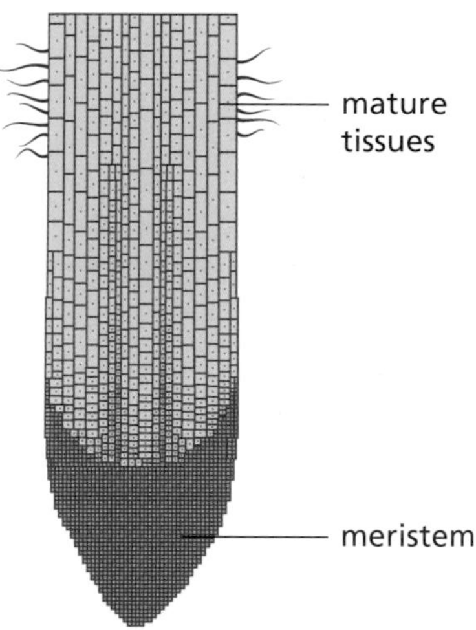

a) Give the function of cells from the meristem tissue. (1)

b) (i) Name the process that results in meristem cells developing into mature tissue in the plant. (1)

(ii) Explain how mature tissues are formed in terms of gene expression. (2)

5 a) Copy and complete the table below to show the differences between embryonic and tissue stem cells. (2)

	Embryonic stem cells	Tissue stem cells
Location	In early embryo tissue	
Potential following cell division		Can differentiate into cells of one type of tissue

b) Describe **one** ethical issue that can arise when considering the use of embryonic stem cells in research or medicine. (2)

6 The phylogenetic tree below shows how several species of carnivorous mammal are related.

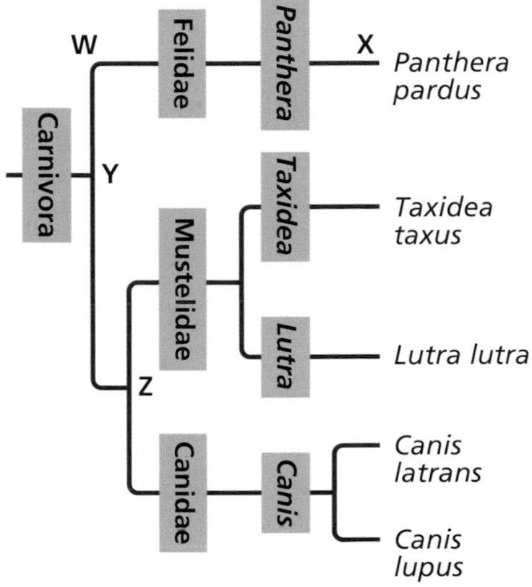

a) Identify the species with DNA sequences most different from *Taxidea taxus*. (1)

b) Identify the species with DNA sequences most similar to *Lutra lutra*. (1)

c) Identify the **two** species whose DNA sequences are **most** similar to each other. (1)

d) Give the letter on the phylogenetic tree that marks the point showing the position of the last common ancestor of the carnivorous mammals. (1)

e) Apart from DNA sequence data, give the other source of information that allows phylogenetic trees to be constructed. (1)

7 In an investigation into the evolution of a group of four closely related species of bird, DNA samples were taken from individuals of each species. The DNA samples were treated with enzymes and the resulting fragments stained, then separated into bands by gel electrophoresis.

The table shows the distances travelled by known sizes of DNA fragments which were used to produce the DNA ladder. The diagram on page 44 shows the results of the gel electrophoresis.

Length of DNA fragment (bp)	Distance travelled in gel (mm)
400	45
550	38
600	20
700	18
900	14
1200	6

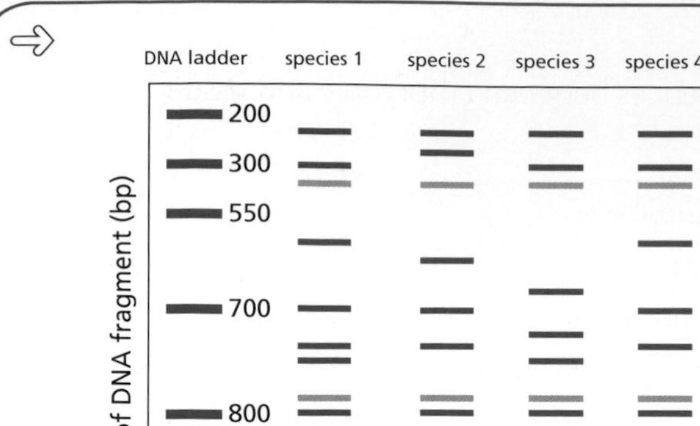

a) Name the types of enzymes that cut DNA at specific points. (1)

b) Identify **one** variable which should be kept constant when treating DNA samples from each species. (1)

c) On a piece of graph paper, draw a line graph to show the distance travelled by the DNA bands against the length of fragments present, as shown in the table. (2)

d) It was concluded that species 1 and 4 were more closely related to each other than to species 2 and 3. Give evidence from the diagram which supports this conclusion. (2)

Question 8 contains a choice.

8 *Either* **A** Give an account of the role of mutation and natural selection in the formation of new species. (7)

Or **B** Give an account of the replication of DNA in cells. (7)

Answers to Practice course assessment: Area 1

Paper 1

1 C, 2 C, 3 A, 4 D, 5 D, 6 A, 7 A, 8 C, 9 D, 10 B [1 each = 10]

Paper 2

1 a) X – phosphate; Y – deoxyribose [2]
b) they run in/are aligned in opposite directions [1]
c) double helix [1]
d) hydrogen bond [1]

2 a) RNA polymerase [1]
b) introns are removed; exons are spliced together [1 each = 2]
c) intron may be left in mRNA *and* wrong amino acids included in polypeptide [1]
d) ribosomes [1]

3 a) AUG [1]
b) threonine *and* proline [1]
c) histidine [1]
d) there would be no difference in the protein produced; both codons code for arginine [1 each = 2]

4 a) can undergo cell division [1]
b) (i) differentiation [1]
(ii) some genes are switched on *and* others are switched off; so that the correct proteins are produced to form the mature tissue [1 each = 2]

5 a) Location: within brain/bone/heart/stomach/named tissue [1]
Potential: can differentiate into any cell type/are pluripotent [1]
b) to access stem cells, embryo would be damaged; medical ethic is to preserve life *or* other possible answers [1 each = 2]

6 a) *Panthera pardus* [1]
b) *Taxidea taxus* [1]
c) *Canis latrans* and *Canis lupus* [1]
d) Y [1]
e) fossils [1]

⇨

7 a) restriction endonuclease [1]

b) mass of sample/temperature/type of enzyme used/type of stain used/others [1]

c) scales and labels with units = 1 correct plotting and plots joined with straight lines = 1 [2]

d) Species 1 and 4 have most bands in common [1]

8 A mutation produces variation within species; mutation can be deleterious or beneficial; there is a struggle for survival; those that are fittest/best suited/with beneficial mutations survive; the survivors breed; survivors pass mutations/favourable or beneficial characteristics to offspring; mutations build up over long periods; new species cannot interbreed to produce fertile offspring [any 7 = 7]

B double helix unwinds; unzips/bond break between complementary bases; primers add onto template strands; primers allow DNA polymerase/enzyme to bind; DNA polymerase adds complementary DNA nucleotide to templates; lead/3′–5′ strand replicated continuously; lagging strand replicated in fragments; fragments joined by ligase; two new identical DNA molecules formed [any 7 = 7]

Area 2 Metabolism and survival

Key Area 2.1a

Metabolic pathways

Key points

1 **Metabolism** is the sum total of all the chemical reactions that take place in cells. ☐
2 A **metabolic pathway** is a series of stepwise chemical reactions that are controlled by enzymes. ☐
3 Metabolic pathways are integrated and controlled pathways of enzyme-catalysed reactions within a cell. ☐
4 Metabolic pathways can have reversible steps, irreversible steps and alternative routes. ☐
5 Reactions within metabolic pathways can be **anabolic** or **catabolic**. ☐
6 Anabolic reactions build up large molecules from small molecules and require energy. ☐
7 Catabolic reactions break down large molecules into smaller molecules and release energy. ☐
8 Proteins embedded in **phospholipid membranes** have functions such as **pores**, **pumps** or enzymes. ☐

Summary notes

Cell metabolism

Metabolism is the name given to the sum total of all the chemical reactions that take place within a living cell. It is the network of connected and integrated pathways involved in cell activities, and involves reversible and irreversible steps, sometimes with alternative routes, that are all controlled by enzymes.

Metabolic pathways

The control of metabolic pathways is essential to cell survival. A metabolic pathway is a series of stepwise chemical reactions controlled by enzymes, as shown in Figure 2.1.

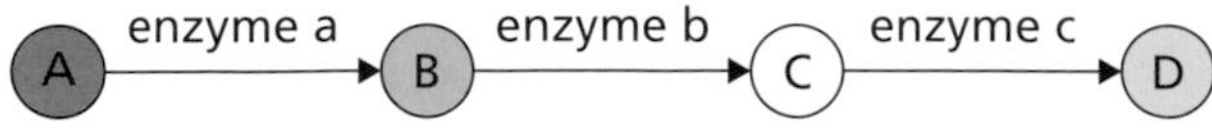

Figure 2.1 Metabolic pathway involving metabolites A–D and enzymes a–c

Anabolism

Anabolic pathways are biosynthetic processes that involve the building up of large molecules from smaller molecules. Anabolic pathways require the input of energy.

Protein synthesis is an example of an anabolic reaction, which requires ATP to provide the energy to build up amino acids to form a protein.

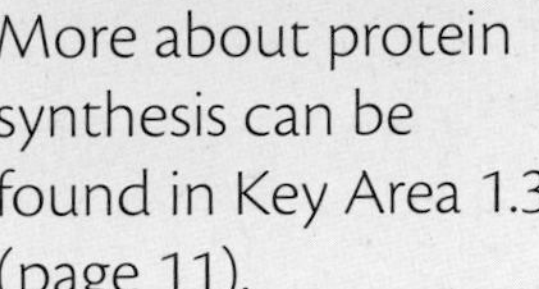

Key links

More about protein synthesis can be found in Key Area 1.3 (page 11).

Catabolism

Catabolic pathways involve the breakdown of larger molecules into smaller molecules. They usually release energy. Catabolism can also provide building blocks for use in other chemical reactions. Aerobic respiration is an example of a catabolic reaction, involving the breakdown of glucose and resulting in the release of energy in the form of ATP.

Key links 

There is more about aerobic respiration in Key Area 2.2 (page 59).

Figure 2.2 shows the two types of metabolic pathway.

Key links 

There is more about ATP in Key Area 2.2 (page 59).

Anabolic reaction
small molecules
+ **energy supplied by ATP**
larger molecules

Catabolic reaction
large molecules
smaller molecules
+ **energy released as ATP**

Figure 2.2 Summary of anabolism and catabolism

Metabolic pathways can have reversible and irreversible steps and alternative routes may exist that can bypass steps in a pathway. Figure 2.3 shows an example of a metabolic pathway, a reversible step, an irreversible step and an alternative bypass route.

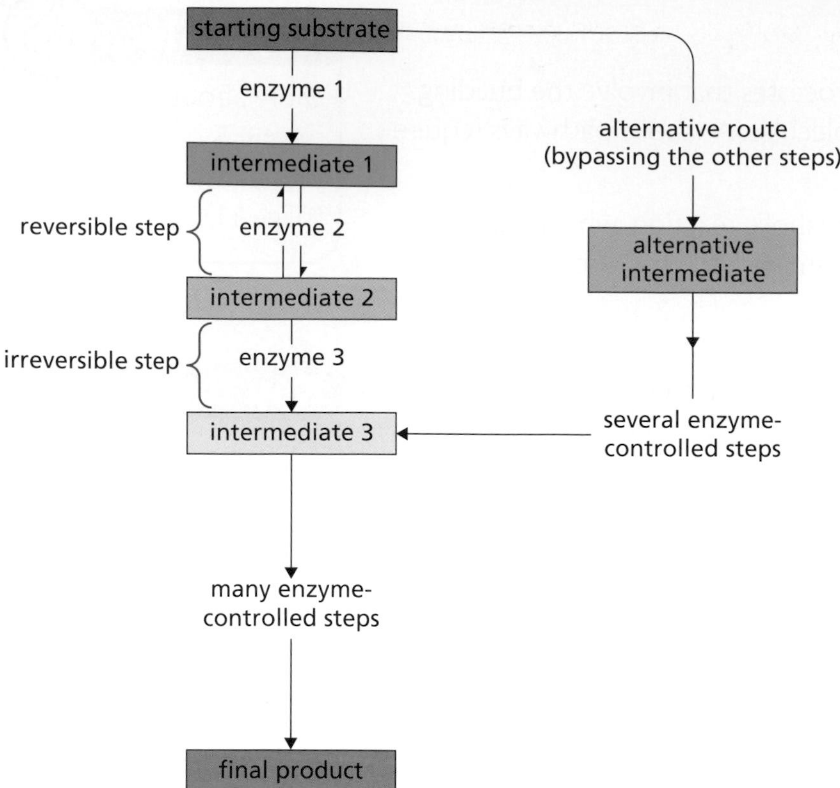

Figure 2.3 A metabolic pathway with characteristic features

Membrane structure

Membranes consist of protein and phospholipid. The phospholipid molecules form a double layer and are in constant motion, giving a fluid nature to membranes and making them flexible.

The proteins are scattered in a patchy mosaic pattern. Some proteins form pores, others are pumps that penetrate through the membrane and some are enzymes that catalyse chemical reactions, as shown in Figure 2.4.

> **Hints & tips**
>
> *Remember the three F sounds – **phospholipid, fluid, flexible**.*

> **Hints & tips**
>
> *Remember **P**oly**PEP**tides – **p**roteins for **p**ores, **e**nzymes and **p**umps.*

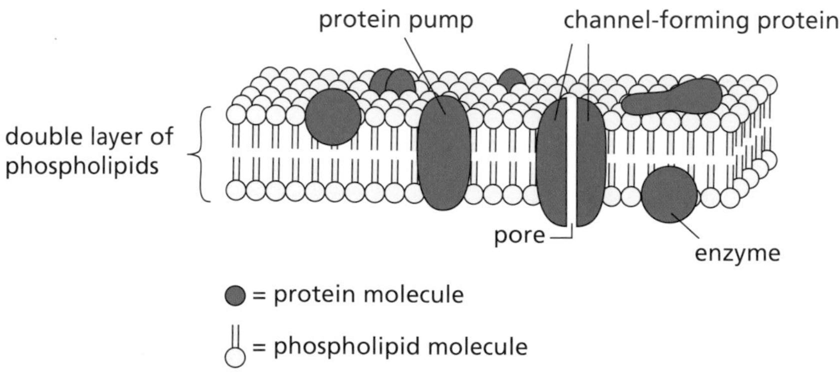

Figure 2.4 The fluid mosaic model of membrane structure

Pores, pumps and enzymes

Channel-forming proteins have pores that control the passive diffusion of certain molecules, depending on their size. These pores make cell membranes selectively permeable.

Active transport involves the movement of molecules from low to high concentration. This type of movement needs additional energy that is available from ATP. Cells must be respiring aerobically to produce enough ATP for active transport. It involves specialised protein pumps in the plasma membrane, which recognise specific molecules and transfer them across the membrane. The sodium–potassium pump is an example of a carrier protein involved in active transport. Figure 2.5 shows how it works in a nerve cell.

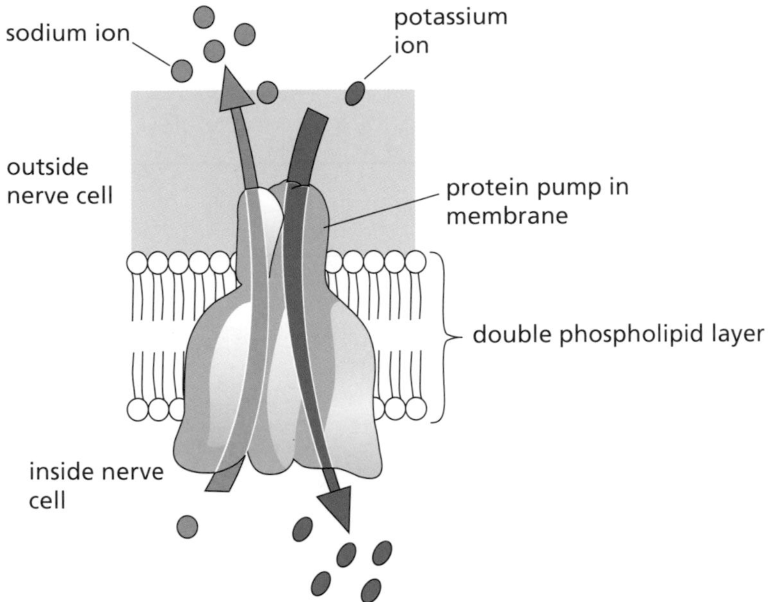

Figure 2.5 The sodium–potassium pump in a nerve cell that pumps ions against their concentration gradients

ATP synthase is an example of an enzyme embedded in the membranes of mitochondria and chloroplasts, where it catalyses the synthesis of ATP, as shown in Figure 2.6.

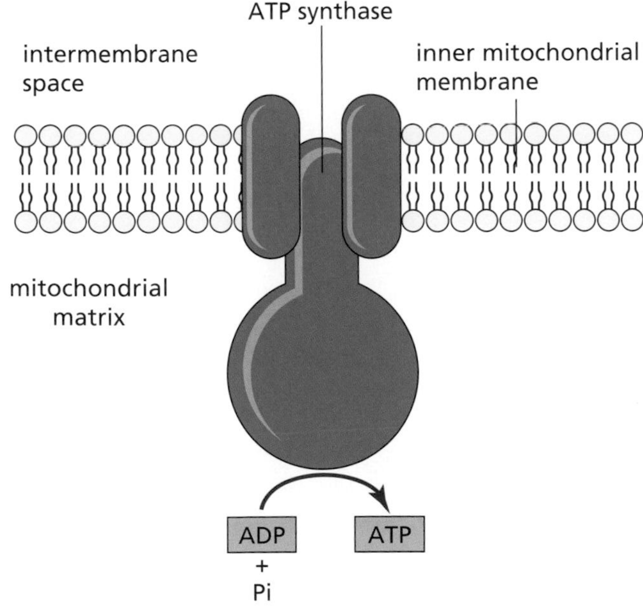

Figure 2.6 ATP synthase embedded in the inner membrane of a mitochondrion

Key links

There is more about ATP synthase in Key Areas 2.2 and 3.1b (pages 59 and 105, respectively).

Key words

Anabolic – metabolic activity that requires energy input and builds up larger molecules from smaller ones
Catabolic – metabolic activity that releases energy from reactions which break down large molecules into smaller ones
Metabolic pathway – enzyme-controlled sequence of chemical reactions in cells
Metabolism – total of all metabolic pathways in an organism
Phospholipid membrane – membrane of a cell made from fluid phospholipid molecules and proteins
Pore – small gap in a membrane created by a channel-forming protein
Pump – protein in a phospholipid membrane that carries substances across it by active transport

Questions ?

Short answer (1 or 2 marks)

1 **a)** Describe what is meant by a metabolic pathway. (1)
b) State **two** differences between anabolic and catabolic pathways. (2)

2 **a)** Give **two** roles of proteins embedded in phospholipid membranes. (2)
b) Give **two** examples of organelles bounded by double membranes. (2)

Longer answer (3–10 marks)

3 Write notes on each of the following:
a) metabolic pathways (4)
b) the structure and function of membranes (4)

Answers are on page 89.

Key Area 2.1b
Control of metabolic pathways

Key points

1 Metabolic pathways are controlled by the presence or absence of particular enzymes and the regulation of the rate of reaction of key enzymes. ☐
2 The **substrate** molecule(s) have a high affinity for the enzyme's **active site** and the subsequent **products** have a low **affinity**, allowing them to leave the active site. ☐
3 **Induced fit** occurs when the active site changes shape to better fit the substrate after the substrate binds. ☐
4 The energy required to initiate a chemical reaction is called the **activation energy**. ☐
5 Enzymes lower the activation energy. ☐
6 The concentrations of substrate and end product affect the direction and rate of an enzyme-controlled reaction. ☐
7 As the substrate concentration increases, the rate of the enzyme reaction increases until all of the active sites are occupied by the substrate. ☐
8 Some metabolic reactions are reversible and the presence of a substrate or the removal of a product will drive a sequence of reactions in a particular direction. ☐
9 The concentration of substrates relative to the concentration of product(s) changes the direction of an enzyme-catalysed reaction. ☐
10 **Competitive inhibition** involves competition for the active site of the enzyme by molecules that resemble the substrate. ☐
11 Competitive inhibitors bind at the active site, preventing the substrate from binding. ☐
12 Competitive inhibition can be reversed by increasing the substrate concentration. ☐
13 **Non-competitive inhibitors** bind away from the active site but change the shape of the active site, preventing the substrate from binding. ☐
14 Non-competitive inhibition cannot be reversed by increasing the substrate concentration. ☐
15 **Feedback inhibition** occurs when the **end product** in the metabolic pathway reaches a critical concentration. ☐
16 The end product inhibits an earlier enzyme, blocking the pathway, and so prevents further synthesis of the end product. ☐

Summary notes
Control of metabolic pathways

Enzymes are coded for by genes. Each step in a metabolic pathway is controlled by a specific enzyme. A metabolic block can occur when a gene mutation results in the absence of a functional enzyme.

Figure 2.7 shows the genetic control of a metabolic pathway, and the result of a block in the pathway.

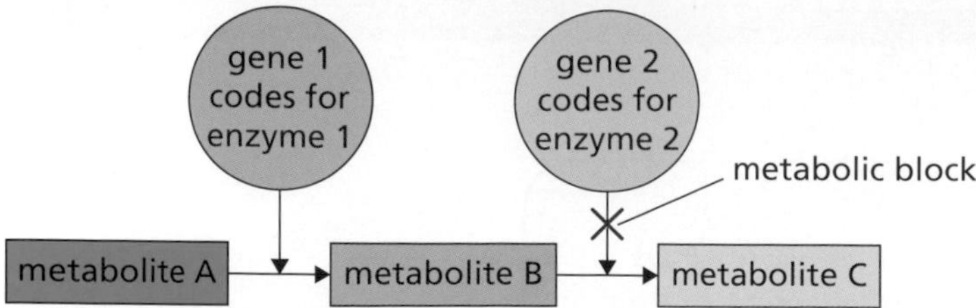

Figure 2.7 Genetic control of a metabolic pathway

If a mutation in gene 2 results in the absence of enzyme 2 then metabolite C will not be produced.

Key links

There is more about gene mutation in Key Areas 1.5 and 1.6 (page 21).

Enzyme action

Enzymes are biological catalysts that speed up the rates of chemical reactions by lowering the activation energy required for the reactions to proceed.

Activation energy

The energy required to initiate a reaction is called its activation energy. Before a substrate can change into a product, the substrate must overcome an energy barrier called the activation energy (E_A).

High temperatures often supply the activation energy in non-living situations, but in cells enzymes reduce the activation energy needed for a reaction to occur, as shown in Figure 2.8.

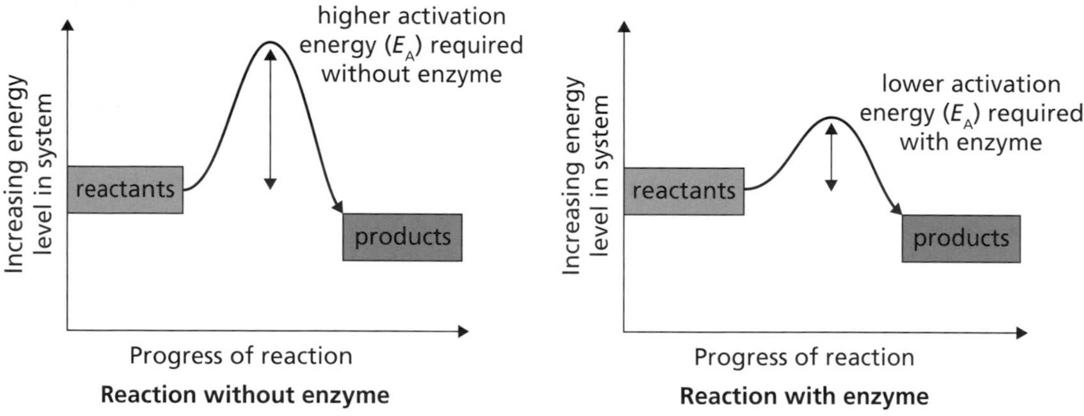

Figure 2.8 Reduction of activation energy due to the presence of a specific enzyme

For example, in the breakdown of hydrogen peroxide:

$E_A = 86\,kJ\,mol^{-1}$ without an enzyme
$E_A = 1\,kJ\,mol^{-1}$ with the enzyme catalase

Induced fit

The active site of an enzyme is the location on its surface where substrate molecules bind and the chemical reaction takes place. Enzymes are specific and only act on one substrate because substrate molecules are complementary in shape to the enzyme's active site. Substrates are chemically attracted to the active site – they are said to have an affinity for it.

As the substrate starts to bind, the active site changes shape to fit the substrate more closely, increasing the rate of reaction, as shown in Figure 2.9.

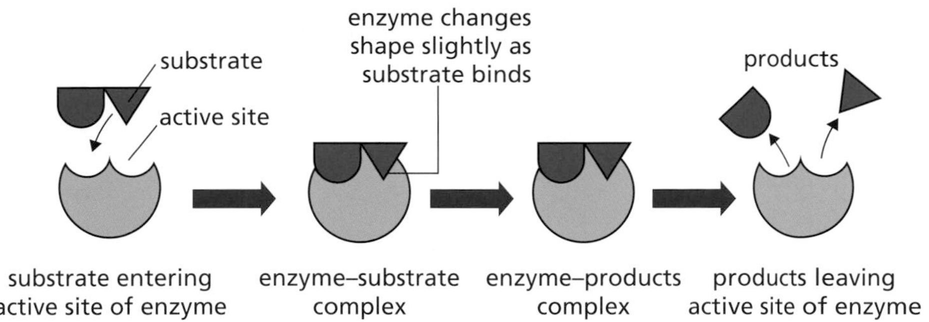

Figure 2.9 The induced-fit model of enzyme action

Hints & tips

Induced fit is a bit like the fitting of a hand into a surgical glove – the glove changes shape and then the fit is exact and very tight.

The active site has high affinity for the substrate molecules and promotes the chemical reaction through the lowering of the activation energy required, as shown in Figure 2.10.

After the reaction takes place, the product, being a different shape from the substrate, moves away because it has low affinity for the active site. The active site returns to its original shape.

Rates of enzyme reaction

The maximum rate at which any enzyme-catalysed reaction can proceed depends on, among other things, the concentration of substrate molecules, as shown in Figure 2.11. An increase in substrate concentration drives the reaction in the direction of the end product and increases the rate of the reaction.

Many metabolic reactions are reversible and the concentration of the substrate and product affect the direction and rate of the reaction.

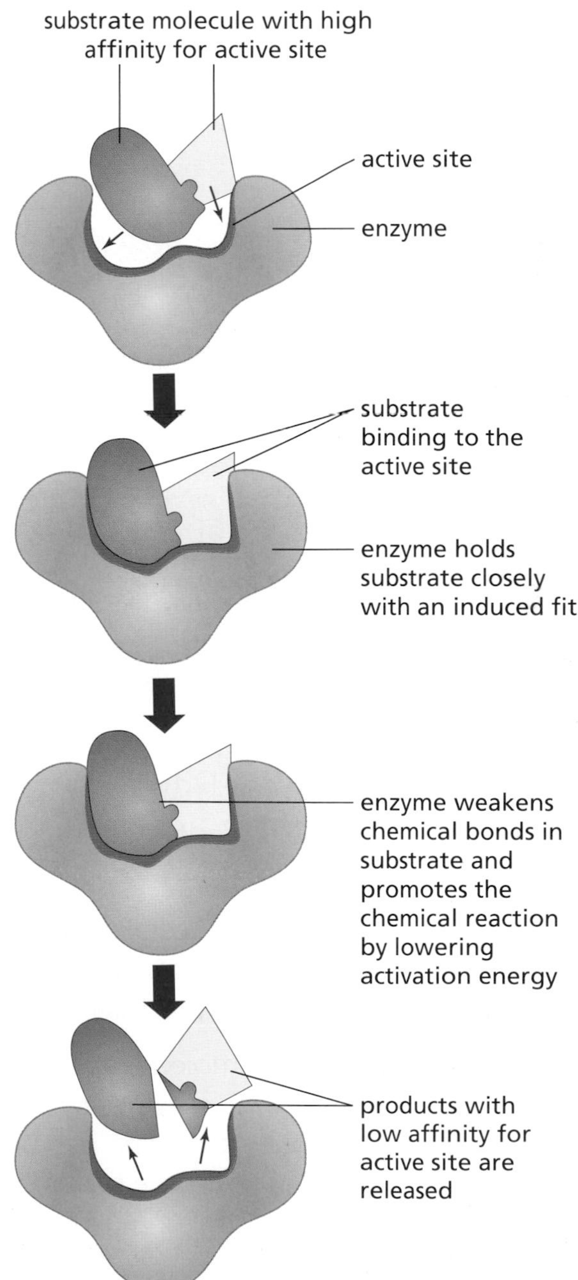

Figure 2.10 The role of the active site in an enzyme-controlled reaction

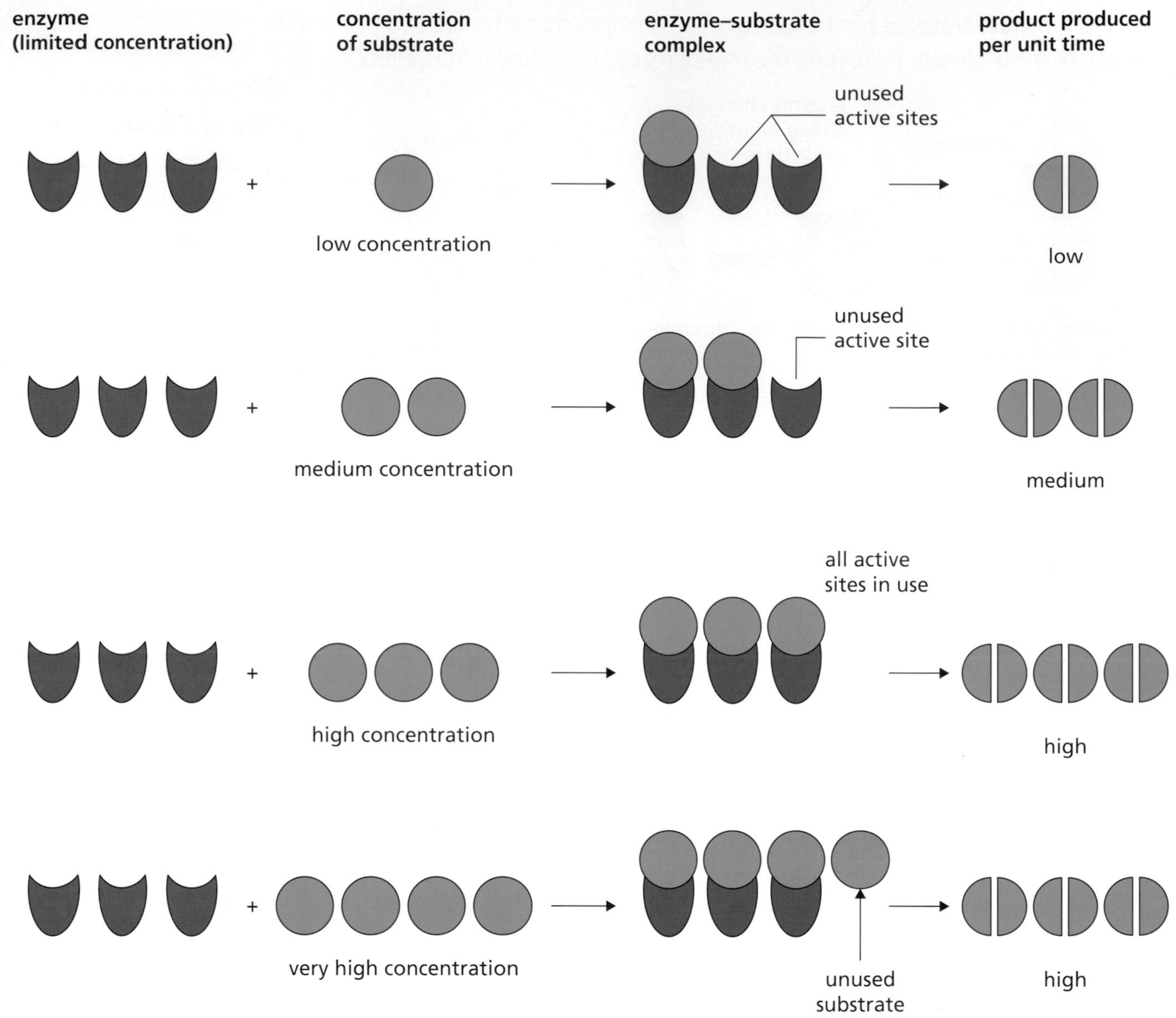

Figure 2.11 Effect of substrate concentration on the rate of an enzyme-catalysed reaction

Control of metabolic pathways through the regulation of enzyme action

Inhibitors

An inhibitor is a substance that reduces the rate of an enzyme reaction. Inhibitors occur naturally but are also produced artificially for uses such as drugs and pesticides. There are two main kinds of inhibitor.

Competitive inhibitors are molecules with a similar structural shape to the normal substrate of the enzyme and so can fit into its active site but do not react to produce products. They compete with substrate molecules for a position in the active site on the enzyme. Figure 2.12 shows the effect of a competitive inhibitor on the activity of an enzyme.

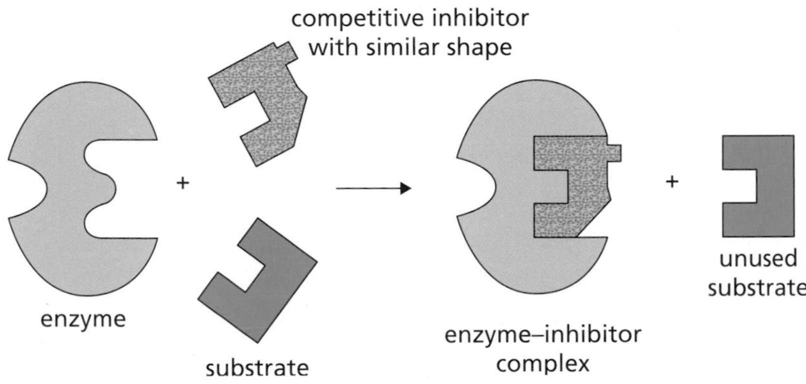

Figure 2.12 Competitive inhibition

With some of the active sites occupied and blocked by the inhibitor, the rate of the reaction is reduced. However, if the substrate concentration is increased, the chance of the substrate binding to the enzyme is increased and the rate of the reaction can return to normal, as shown in Figure 2.13.

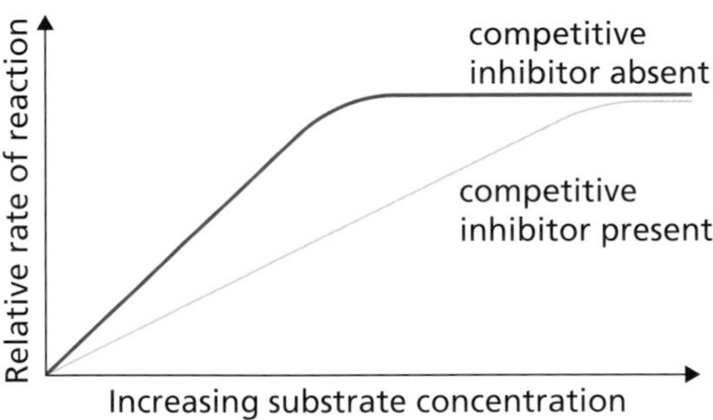

Figure 2.13 The effect of increasing substrate concentration on competitive inhibition

Non-competitive inhibitors are molecules with a quite different structure from the substrate molecule. They do not fit into the active site of the enzyme but bind to another part of the enzyme molecule. This changes the shape of the active site, so that it can no longer combine with the substrate molecule. Figure 2.14 shows the effect of a non-competitive inhibitor on the activity of an enzyme.

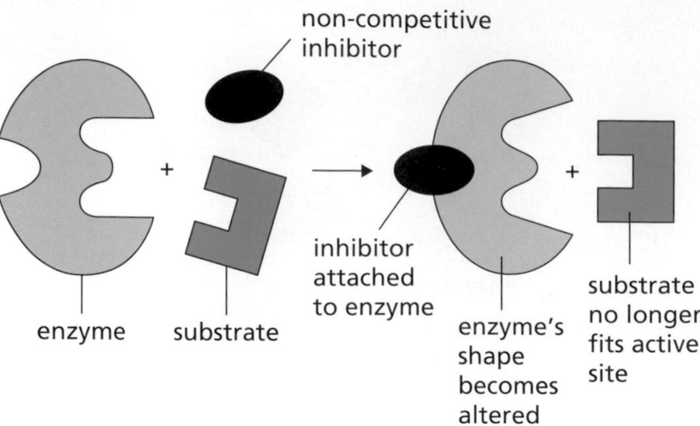

Figure 2.14 Non-competitive inhibition

The effect of non-competitive inhibitors is to reduce the amount of active enzyme and has a similar effect to decreasing the enzyme concentration. Cyanide, heavy metal ions and some insecticides are examples of non-competitive inhibitors. Increasing the substrate concentration does not increase the reaction rate in the presence of a non-competitive inhibitor and the effect of the inhibitor is permanent.

Feedback inhibition occurs when an end product inhibits the activity of an enzyme that catalysed a reaction earlier in the pathway that produced it, as shown in Figure 2.15.

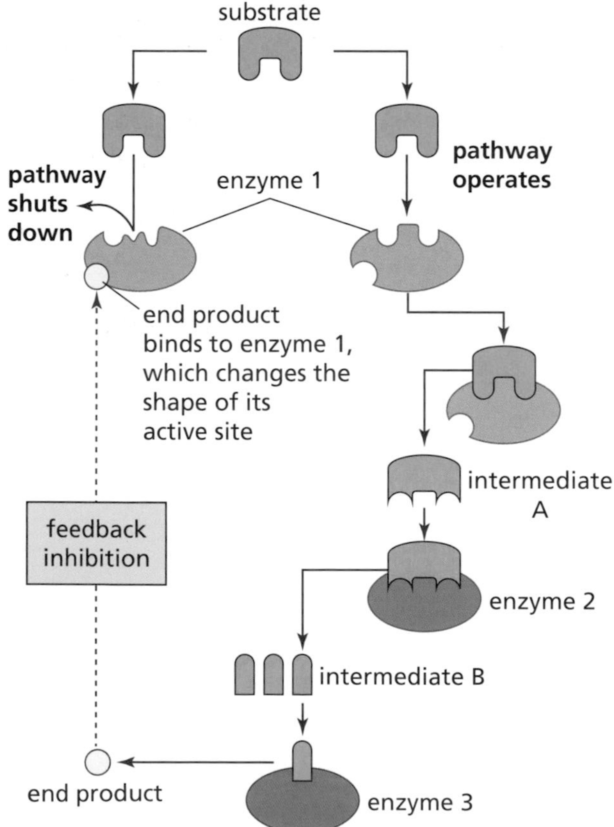

Figure 2.15 Feedback inhibition

Hints & tips

Feedback inhibition can be called end product inhibition.

The effect of substrate or inhibitor concentration on the rate of enzyme-controlled reactions is something you need to be familiar with for your exam.

Figure 2.16(a) shows the results of an experiment in which an enzyme reaction has been carried out at various substrate concentrations. As the substrate concentration is increased from 0 to X the rate of reaction increases as more active sites are gradually being used. Between X and Y the reaction rate does not increase as the active sites are always occupied and a maximum rate has been achieved.

Figure 2.16(b) shows the results of an experiment in which an enzyme reaction has been carried out at various inhibitor concentrations. As the inhibitor concentration is increased from 0 to X the rate of reaction decreases as more active sites are occupied by inhibitor. Between X and Y the reaction has almost stopped because active sites are always occupied by inhibitor.

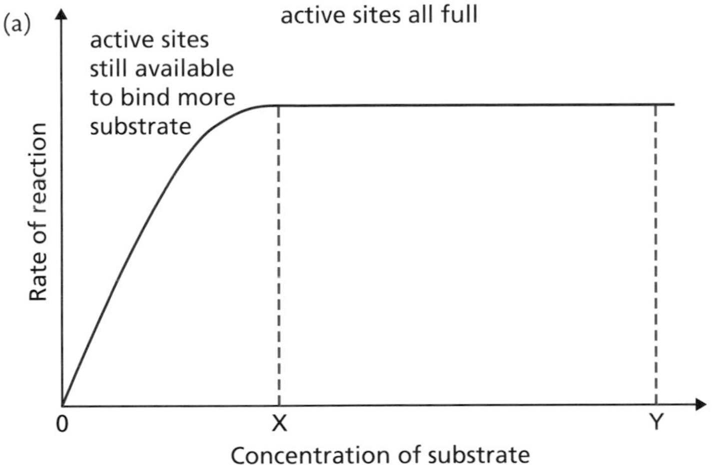

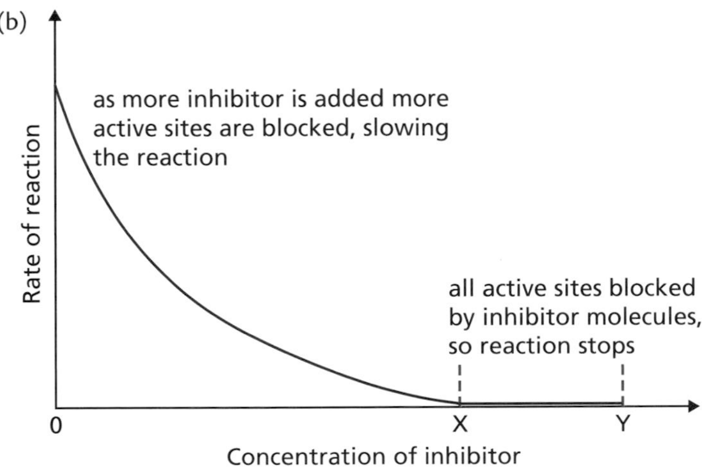

Figure 2.16 Effect of **(a)** substrate and **(b)** inhibitor concentrations on the rate of an enzyme-controlled reaction

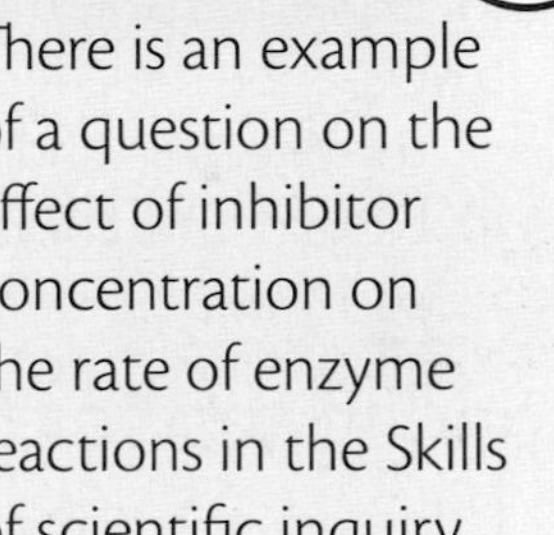

There is an example of a question on the effect of inhibitor concentration on the rate of enzyme reactions in the Skills of scientific inquiry chapter on page 146.

Key words

Activation energy – input of energy required to start a chemical reaction
Active site – region on an enzyme molecule where the substrate binds
Affinity – the degree to which molecules are likely to combine with each other
Competitive inhibition – the slowing of reaction rate due to the presence of a substance resembling the substrate
End product – a final product in a metabolic reaction which is not further converted to other substances
Feedback inhibition – enzyme inhibition caused by the presence of an end product of a metabolic pathway acting as an inhibitor of an enzyme earlier in the pathway
Induced fit – change to an enzyme's active site brought about by its substrate
Non-competitive inhibition – enzyme inhibition by a substance that permanently alters the active site of an enzyme
Product – substance resulting from an enzyme-catalysed reaction
Substrate – substance on which an enzyme acts

Questions ?

Short answer (1 or 2 marks)

1 Describe the role of genes in the control of metabolic pathways. (2)

2 a) Explain what is meant by the induced-fit model of enzyme action. (2)
b) Describe the effect of an increase in substrate concentration on the direction and rate of an enzyme reaction. (2)
c) Explain how enzymes speed up the rate of reactions in metabolic pathways. (2)

Longer answer (3–10 marks)

T

3 Using Figure 2.16 (b), explain the effect on the rate of an enzyme reaction of increasing the concentration of an inhibitor of that enzyme. (4)

4 Give an account of enzyme action and of the effects of competitive and non-competitive inhibition. (9)

Answers are on page 89.

Key Area 2.2
Cellular respiration

Key points

1 **Glycolysis** is the breakdown of glucose to **pyruvate** in the cytoplasm. ☐
2 **ATP** is required for the **phosphorylation** of glucose and **intermediates** during the **energy investment phase** of glycolysis. This leads to the generation of more ATP during the **energy pay-off stage** and results in a **net gain** of ATP. ☐
3 Pyruvate progresses to the **citric acid cycle** if oxygen is available. ☐
4 In aerobic conditions, pyruvate is broken down to an **acetyl group** that combines with **coenzyme A** forming acetyl coenzyme A. ☐
5 In the citric acid cycle the acetyl group from acetyl coenzyme A combines with **oxaloacetate** to form **citrate**. ☐
6 During a series of enzyme-controlled steps, citrate produced in the citric acid cycle is gradually converted back into oxaloacetate which results in the generation of ATP and release of carbon dioxide. ☐
7 The citric acid cycle occurs in the **matrix** of the mitochondria. ☐
8 In both glycolysis and the citric acid cycle, **dehydrogenase** enzymes remove **hydrogen ions** and **electrons** and pass them to the coenzyme **NAD**, forming NADH. ☐
9 Hydrogen ions and electrons from NADH are passed to the **electron transport chain** on the inner mitochondrial membrane. ☐
10 The electron transport chain is a series of carrier proteins attached to the inner mitochondrial membrane. ☐
11 Electrons are passed along the electron transport chain, releasing the energy which allows hydrogen ions to be pumped across the inner mitochondrial membrane. ☐
12 The return flow of hydrogen ions back through the membrane protein **ATP synthase** results in the synthesis of ATP. ☐
13 Hydrogen ions and electrons combine with oxygen to form water. ☐
14 In the absence of oxygen, **fermentation** takes place in the cytoplasm. ☐
15 In animal cells, pyruvate is converted to **lactate** in a reversible reaction. ☐
16 In plants and yeast, ethanol and CO_2 are produced in an irreversible reaction. ☐
17 Fermentation results in much less ATP being produced than in aerobic respiration. ☐
18 ATP is used to transfer energy from **cellular respiration** to synthesis processes and other cellular processes which require energy. ☐

Summary notes

Cellular respiration

Cellular respiration pathways are present in the cells from all three domains of life.

Key links

There is more about domains of life in Key Area 1.8 (page 32).

The metabolic pathways of cellular respiration are of central importance to cells. They yield energy and are connected to many other pathways.

Transfer of energy via ATP

ATP is built up or regenerated from ADP and **inorganic phosphate (Pi)** using the energy released from cellular respiration. Respiration is a catabolic pathway which converts the chemical energy stored in glucose into chemical energy stored in ATP.

ATP is used to transfer the chemical energy from cellular respiration to anabolic synthesis pathways and other cellular processes where energy is required, for example the contraction of muscle fibres, active transport, DNA replication and protein synthesis. The energy held in the ATP is released when it is broken down into ADP + Pi. Figure 2.17 shows an example of the transfer of chemical energy by ATP to a synthetic pathway.

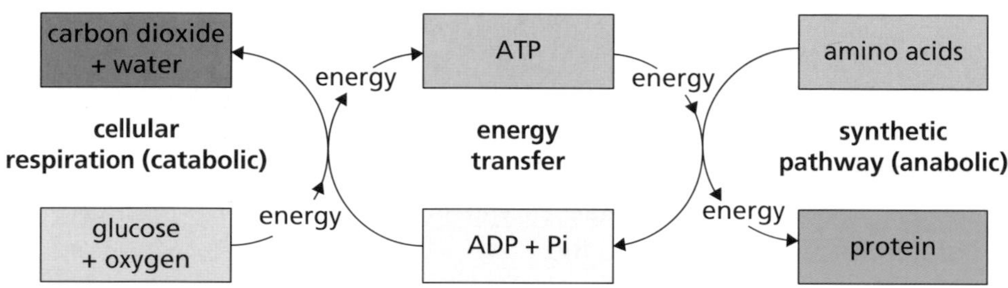

Figure 2.17 An example of the role of ATP in energy transfer

Synthesis of ATP in respiration

During respiration, glucose is broken down in a series of enzyme-catalysed steps. Hydrogen ions and electrons are removed by dehydrogenase enzymes and used in the synthesis of ATP.

Stages in aerobic respiration

The stages of respiration take place in different parts of the cell.

1 Glycolysis

Glycolysis is a series of enzyme-controlled reactions that take place in the cytoplasm of cells.

During glycolysis glucose is broken down to pyruvate in the absence of oxygen.

The phosphorylation of intermediates in glycolysis uses two molecules of ATP and is described as an energy investment phase. The later reactions in glycolysis result in the direct regeneration of four molecules of ATP for every glucose molecule and are referred to as the energy pay-off phase, giving a net gain of 2 ATP. During the energy pay-off phase, dehydrogenase enzymes remove hydrogen ions (H^+) and electrons, which combine with the coenzyme NAD to form NADH. If oxygen is present, NADH transports hydrogen to the electron transport chain, which leads to the production of more ATP. Figure 2.18 shows the breakdown of glucose to pyruvate during glycolysis.

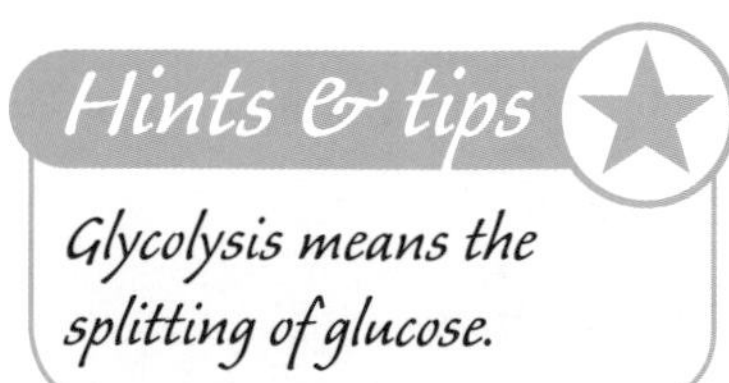

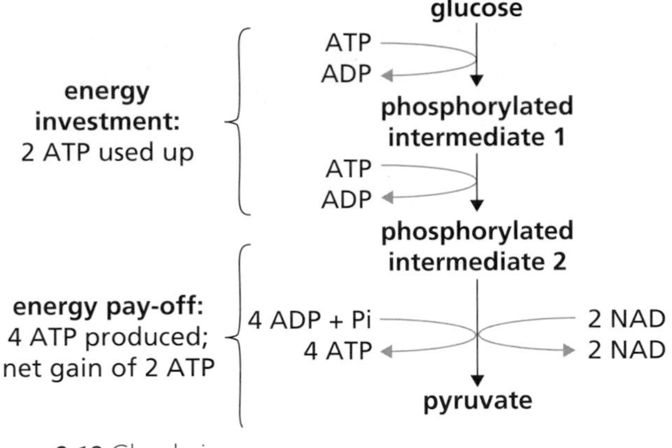

Figure 2.18 Glycolysis

2 Citric acid cycle

If oxygen is available, pyruvate progresses to the citric acid cycle. This stage takes place in the central matrix of the mitochondria. Pyruvate enters the mitochondria and is broken down by enzymes to an acetyl group and carbon dioxide (CO_2).

The acetyl group then combines with coenzyme A to be transferred to the citric acid cycle as acetyl coenzyme A. The acetyl group combines with oxaloacetate to form citrate (citric acid). The citrate then undergoes a series of enzyme-mediated steps resulting in the generation of one ATP molecule and the release of carbon dioxide. Citrate is gradually converted back to oxaloacetate in the matrix of the mitochondria. During the citric acid cycle, dehydrogenase enzymes remove hydrogen ions (H^+) and electrons, which combine with the coenzyme NAD to form NADH. Figure 2.19 shows the stages involved in the citric acid cycle.

In the absence of oxygen, the pyruvate undergoes fermentation to either lactate in animal cells or ethanol and CO_2 in plant cells and in yeast as shown in Figure 2.20.

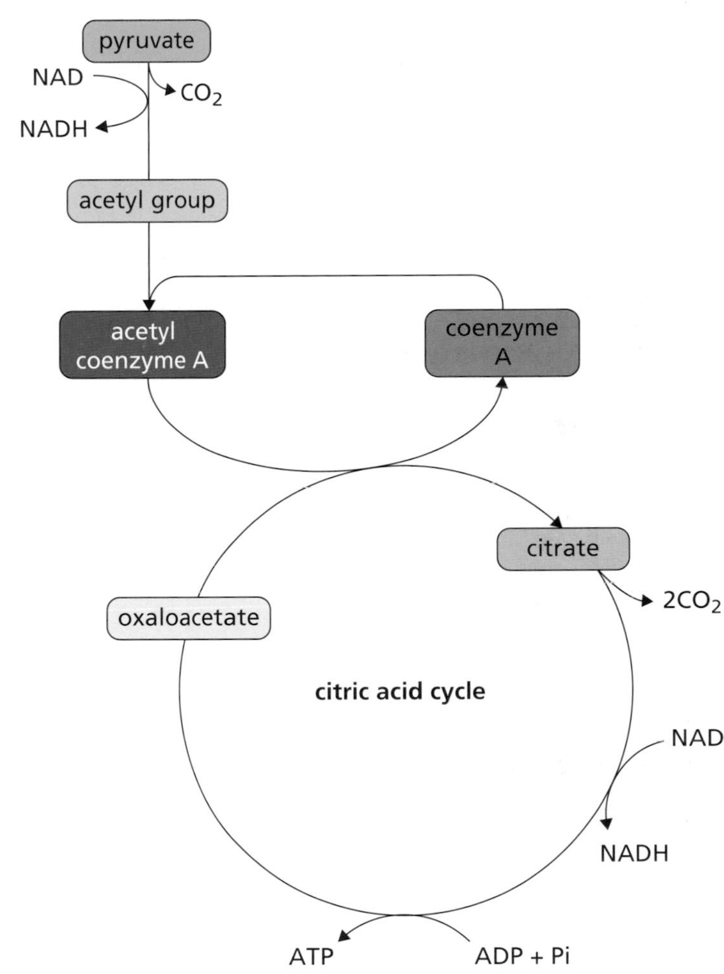

Figure 2.19 Citric acid cycle

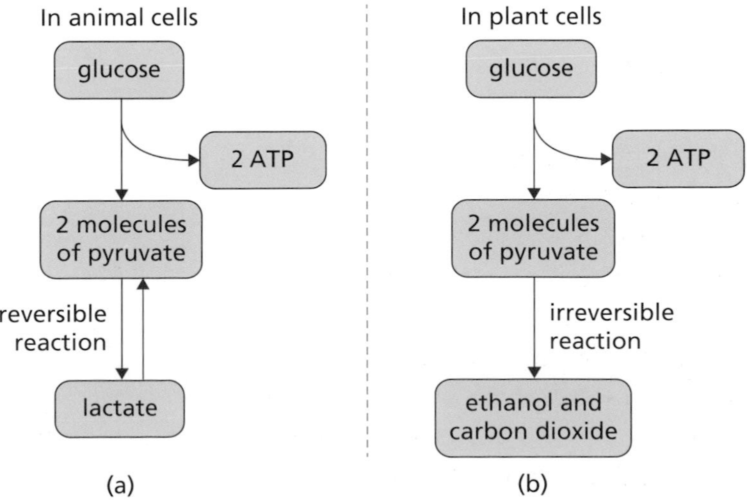

Figure 2.20 Fermentation in (a) animal cells and (b) plant and yeast cells

3 Electron transport chain

NADH transports and passes on hydrogen ions (H^+) and electrons to the electron transport chain. This final stage in aerobic respiration takes place on the inner membrane of the mitochondria. The electron transport chain is a collection of proteins attached to the membrane.

NADH releases the electrons to the electron transport chain where they pass down the chain of electron acceptors, releasing their energy. The energy is used to pump hydrogen ions (H^+) across the inner mitochondrial membrane from the matrix side of the mitochondria into the space between its membranes. The return flow of the hydrogen ions (H^+) back into the matrix drives the enzyme ATP synthase, which results in the synthesis of ATP from ADP + Pi. This stage produces most of the ATP generated by cellular respiration. Oxygen is the final acceptor of hydrogen ions and electrons, forming water.

Key links 

There is more about proteins in membranes in Key Area 2.1a (page 46).

Figure 2.21 shows the stages involved in the electron transport chain on the inner membranes of mitochondria.

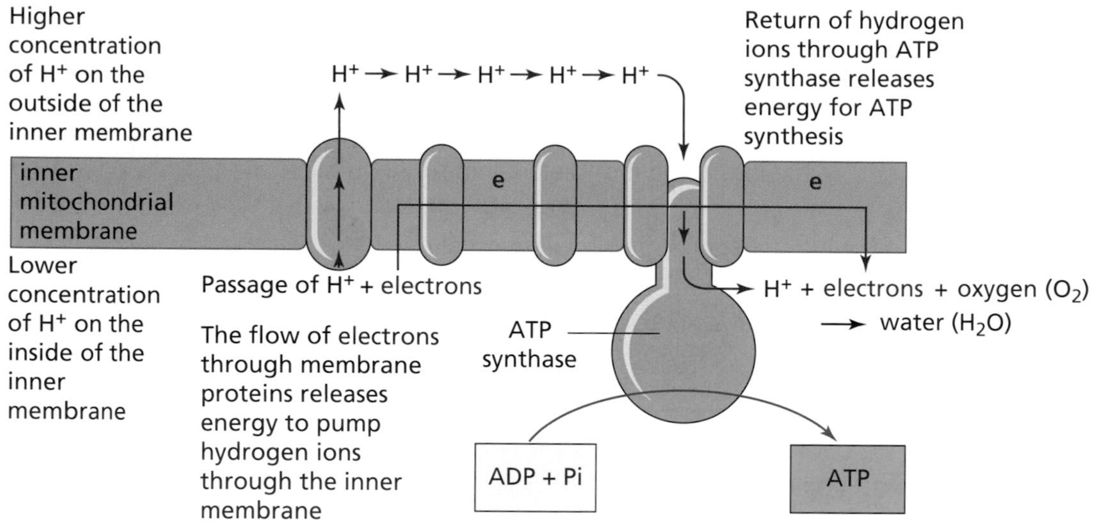

Figure 2.21 The electron transport chain of respiration on the inner membranes of mitochondria

Key words

Acetyl group – produced by breakdown of pyruvate; joins with oxaloacetate in the citric acid cycle
ATP – molecule used for energy transfer in cells
ATP synthase – enzyme within a phospholipid membrane which produces ATP from ADP + Pi
Cellular respiration – release of energy from respiratory substrates
Citrate – citric acid; first substance produced in the citric acid cycle
Citric acid cycle – second stage of aerobic respiration, occurring in the matrix of mitochondria
Coenzyme A – substance that carries an acetyl group into the citric acid cycle as acetyl coenzyme A
Dehydrogenase – enzyme which removes hydrogen from its substrate; important in the citric acid cycle
Electron – negatively charged particle that yields energy as it passes through an electron transport chain
Electron transport chain – group of proteins embedded in membranes of mitochondria and chloroplasts
Energy investment phase – stage of glycolysis in which ATP is used up
Energy pay-off stage – stage of glycolysis in which a net gain of ATP is made
Fermentation – progression of pyruvate in the absence of oxygen
Glycolysis – first stage in cellular respiration which occurs in the cytoplasm
Hydrogen ion (H^+) – a hydrogen atom which has lost an electron leaving it positively charged
Inorganic phosphate (Pi) – used to phosphorylate ADP
Intermediate – substance in a metabolic pathway between the original substrate and the end product
Lactate – produced by the anaerobic conversion of pyruvate in mammalian muscle cells
Matrix – central fluid-filled cavity of a mitochondrion
NAD – coenzyme which carries hydrogen and electrons from glycolysis and the citric acid cycle to the electron transport chain
Net gain – overall production of ATP is greater than ATP used up
Oxaloacetate – substance that combines with the acetyl group in the citric acid cycle to form citrate
Phosphorylation – addition of phosphate to a substance
Pyruvate – end product of glycolysis

Questions ?

Short answer (1 or 2 marks)

1 a) Explain why the phosphorylation of intermediates in glycolysis is described as an energy investment phase. (2)
b) State the role of dehydrogenase enzymes in glycolysis and the citric acid cycle. (1)
c) Describe the role of the coenzymes NAD. (2)

2 a) Name the enzyme embedded in the inner membrane of a mitochondrion responsible for the regeneration of ATP. (1)
b) Describe the role of the electrons transported to the electron transport chain. (2)
c) State the role of oxygen in the electron transport chain. (1)

Longer answer (3–10 marks)

3 Give an account of glycolysis and the citric acid cycle in respiration. (9)
4 Give an account of the electron transport chain and the transfer of energy by ATP. (9)

Answers are on page 90.

Key Area 2.3

Metabolic rate

Key points

1. The **metabolic rate** of an organism is the amount of energy used in a given period of time. ☐
2. The metabolic rates of different organisms at rest can be compared through the measurement of oxygen consumption, carbon dioxide production and heat production. ☐
3. Metabolic rate can be measured using **respirometers**, **oxygen probes**, **carbon dioxide probes** and **calorimeters**. ☐
4. Organisms with high metabolic rates require more efficient delivery of oxygen to cells. ☐
5. Birds and mammals have higher metabolic rates than reptiles and amphibians, which in turn have higher metabolic rates than fish. ☐
6. In **double circulation**, blood passes through the heart twice during a full circulation of the body.
7. Birds and mammals have a **complete double circulatory** system consisting of two **atria** and two **ventricles**. ☐
8. Amphibians and most reptiles have an **incomplete double circulatory** system consisting of two atria and one ventricle. ☐
9. Fish have a **single circulatory system** consisting of one atrium and one ventricle. ☐
10. Complete double circulatory systems, such as those found in birds and mammals, enable higher metabolic rates to be maintained. ☐
11. In complete double systems there is no mixing of oxygenated and deoxygenated blood, and the oxygenated blood can be pumped out at a higher pressure enabling more efficient oxygen delivery to cells. ☐

Summary notes

Metabolic rate

The metabolic rate of an organism is the amount of energy used in a given period of time and can be measured indirectly in terms of the oxygen consumed or the carbon dioxide produced and directly by the heat produced in the time period.

T Measuring metabolic rate is a technique you need to be familiar with for your exam. Metabolic rates can be measured indirectly by using respirometers, oxygen probes, carbon dioxide probes and directly using calorimeters. Respirometers usually measure the consumption of oxygen by organisms over a period of time, as shown by the example in Figure 2.22(a). Oxygen and carbon dioxide probes can give information on the changes to gas levels caused by organisms in a closed container, as shown in Figure 2.22(b). Calorimeters can directly measure the heat produced by organisms over a period of time, as shown in Figure 2.22(c).

Key links

There is an example of a question on metabolic rates in the Skills of scientific inquiry chapter on page 146.

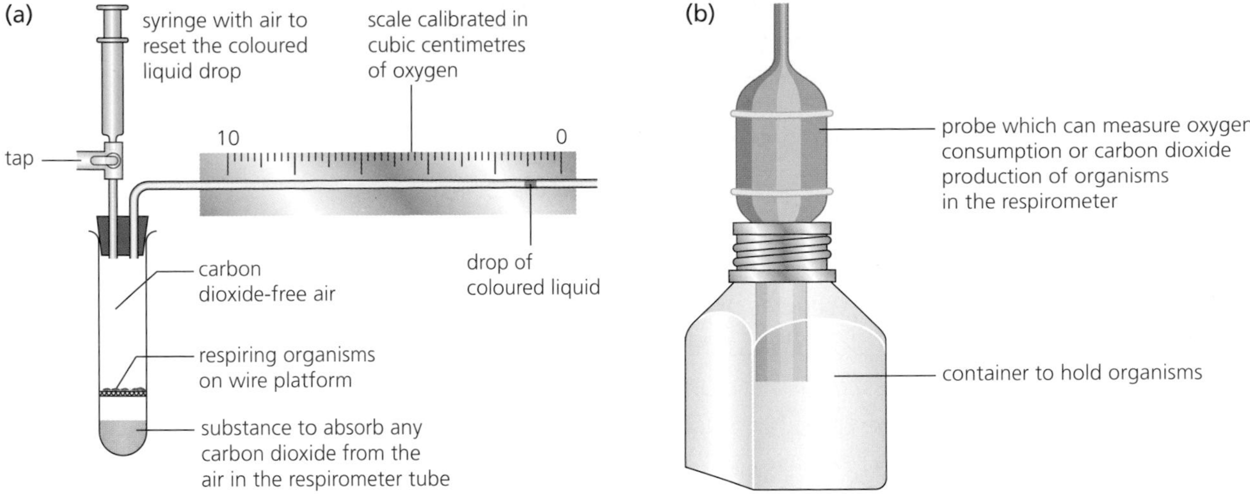

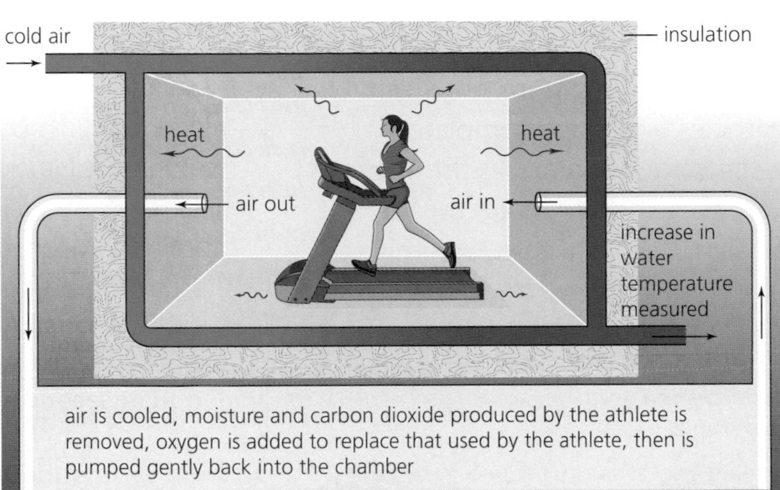

Figure 2.22(a) A simple respirometer to measure the oxygen consumption by organisms, **(b)** gas probe set up to measure changes in gas levels in a respirometer holding organisms and **(c)** calorimeter to measure heat production by a human

Respirometry is a technique which you need to become familiar with for your exam.

Key links 

There is an example of a question on using a respirometer in the Skills of scientific inquiry chapter on page 146.

Oxygen delivery

Oxygen is consumed during aerobic respiration, and organisms that have high metabolic rates require the efficient delivery of oxygen to their cells. In vertebrates, oxygen is delivered in blood pumped by the heart. Vertebrate hearts and circulatory systems have evolved to increase efficiency of the delivery of oxygen to tissues.

Circulatory systems in vertebrates

1 Single circulatory system

Fish have a single circulatory system, which has a heart with two chambers – an atrium and a ventricle. It is called a single circulatory system because the blood only passes through the heart once in each complete circuit, as shown in Figure 2.23.

When blood passes through a capillary bed a drop in pressure occurs. In fish, this means that blood is delivered to the capillary bed in the body tissues at low pressure, which can be inefficient.

2 Incomplete double circulatory system

Amphibians and most reptiles have incomplete double circulatory systems. Their hearts have three chambers – right and left atria and a ventricle. The right atrium receives deoxygenated blood returning from the capillary bed in the body tissues. The left atrium receives oxygenated blood returning from the lungs, which allows the blood pressure to be maintained. The blood from both atria is then passed into the one ventricle, which means that the oxygenated and deoxygenated blood mix before being pumped out of the ventricle to supply the body tissues with the oxygen they require, as shown in Figure 2.24. The blood delivered to the tissues is a mix of oxygenated and deoxygenated, making the system inefficient.

3 Complete double circulatory system

Birds and mammals have a complete double circulatory system. The heart has four chambers – two atria and two ventricles – allowing complete separation of oxygenated and deoxygenated blood. Blood passes through the heart twice during each complete circuit of the body with no mixing of the oxygenated and deoxygenated blood, as shown in Figure 2.25. This is efficient since the blood delivered to tissues is fully oxygenated and at high pressure.

Hints & tips

Reptiles such as crocodile and their relatives have evolved four-chambered hearts as an adaptation to maintaining higher metabolic rates.

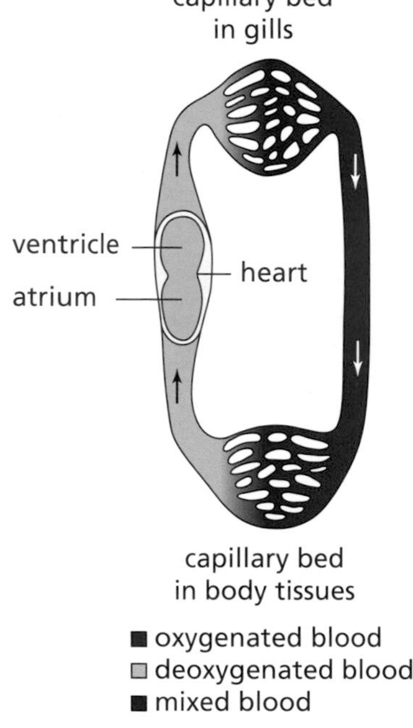

Figure 2.23 Fish circulation

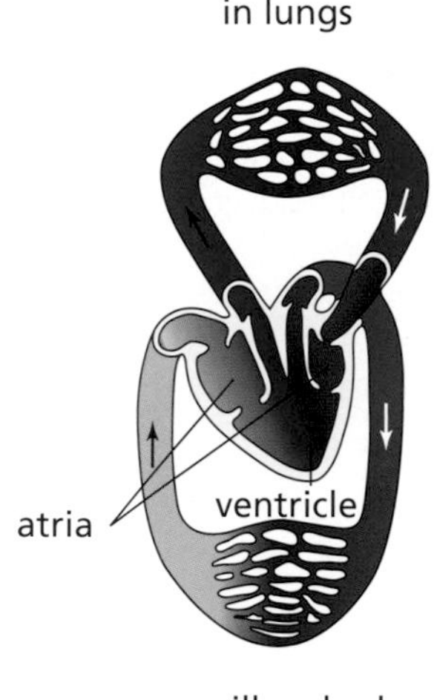

Figure 2.24 Circulation in amphibians and **most** reptiles

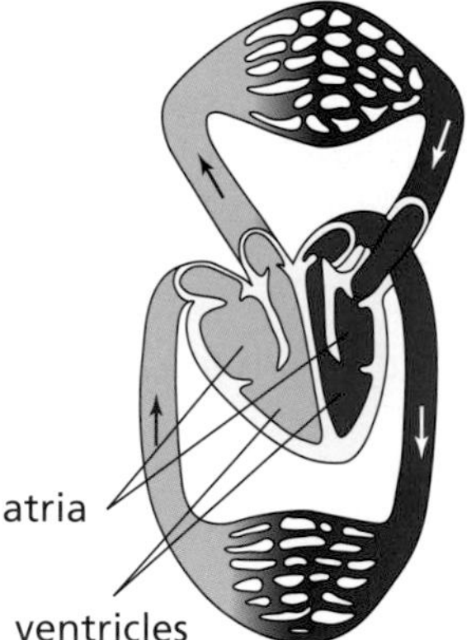

Figure 2.25 Bird and mammal circulation

The various circulatory systems are compared in the following table.

Vertebrate group(s)	Type of circulation	Heart	Features
Fish	Single	Two chambers – an atrium and a ventricle	Loss of blood pressure reduces efficiency
Amphibians and most reptiles	Incomplete double	Three chambers – a right and left atrium and a ventricle	Pressure maintained, but efficiency reduced because tissue blood is incompletely oxygenated
Birds and mammals	Complete double	Four chambers – two atria and two ventricles	Pressure maintained and tissue blood completely oxygenated making the system very efficient

Key words

Atria – chambers receiving blood entering a vertebrate heart (singular: atrium)
Calorimeter – device for measuring heat production by organisms
Carbon dioxide and oxygen probes – devices for measuring the production of carbon dioxide and the uptake of oxygen by organisms
Complete double circulation – double circulation with complete separation of oxygenated and deoxygenated blood (e.g. in birds and mammals)
Double circulation – blood flows through the heart twice during a full circulation of the body
Incomplete double circulation – double circulation with some mixing of oxygenated and deoxygenated blood (e.g. in amphibians and some reptiles)
Metabolic rate – rate of consumption of energy by an organism
Respirometer – device for measuring the oxygen consumption of organisms
Single circulatory system – blood flows through the heart once during a full circulation of the body (e.g. in fish)
Ventricle – chamber of a vertebrate heart that distributes blood

Questions ?

Short answer (1 or 2 marks)

1 **a)** State what is meant by an organism's metabolic rate. (1)
b) Describe **two** methods of measuring the metabolic rate of an organism. (2)
2 State what is measured by the following pieces of apparatus:
a) respirometer (1)
b) calorimeter. (1)

Longer answer (3–10 marks)

3 Compare and contrast the heart structure and circulation of fish, amphibians and mammals. (9)

Answers are on page 91.

Key Area 2.4

Metabolism in conformers and regulators

Key points !

1 The ability of an organism to maintain its metabolic rate is affected by external abiotic factors. ☐

2 Abiotic factors that can affect the metabolic rate of an organism include temperature, salinity and pH. ☐

3 The internal environment of a **conformer** is dependent on its external environment. ☐

4 Conformers use behavioural responses to maintain their optimum metabolic rate. ☐

5 Behavioural responses by conformers allow them to tolerate variation in their external environment to maintain an optimum metabolic rate. ☐

6 Conformers have low metabolic costs and a narrow range of **ecological niches**. ☐

7 Homeostasis is the maintenance of steady conditions within an organism. ☐

8 **Regulators** maintain their internal environment regardless of the external environment. ☐

9 Regulators use metabolism to control their internal environment, which increases the range of possible ecological niches they can occupy. ☐

10 Regulators require energy to achieve **homeostasis** and as a result have high metabolic costs. ☐

11 **Thermoregulation** is the control of body temperature by **negative feedback**. ☐

12 Negative feedback is the control mechanism by which homeostasis is achieved. ☐

13 Negative feedback systems have monitoring centres with receptor cells, a system for sending messages and effectors, which carry out a response. ☐

14 The **hypothalamus** is the temperature-monitoring centre of the mammalian brain and contains **thermoreceptors**, which detect changes in blood temperature. ☐

15 Information is communicated by electrical impulses through nerves to the effectors, which bring about corrective responses to return temperature to normal. ☐

16 Corrective responses to an increase in body temperature include increased sweating, **vasodilation** of blood vessels and decreased metabolic rate. ☐

17 With increased sweating, more body heat is used to evaporate water in the sweat, cooling the skin. ☐

18 Vasodilation increases blood flow to the skin, increasing heat loss. ☐

19 When metabolic rate decreases less heat is produced. ☐

⇨

➩

20 Corrective responses to a decrease in body temperature include shivering, **vasoconstriction** of blood vessels, contraction of hair erector muscles and increased metabolic rate. ☐

21 Shivering is involuntary muscle contraction, which generates heat to replace the heat lost. ☐

22 Vasoconstriction decreases blood flow to the skin decreasing heat loss. ☐

23 Hair erector muscles contract to raise hair trapping a layer of insulating air. ☐

24 When metabolic rate increases more heat is produced. ☐

25 Thermoregulation is important for optimal enzyme-controlled reaction rates and high diffusion rates for the maintenance of metabolism.

Summary notes

Metabolism in conformers and regulators

External abiotic factors such as temperature, salinity and pH can affect the ability of an organism to maintain its metabolic rate.

Conformers

Organisms that cannot maintain their metabolic rate by physiological mechanisms are called conformers. A conformer's internal environment is directly dependent on its external environment. They do not use physiological mechanisms that require energy to alter their metabolic rate and so they have low metabolic energy costs. Conformers use behavioural responses to maintain optimum metabolic rate and have a narrow range of ecological niches.

Behavioural responses by conformers, such as reptiles basking on rocks to increase body temperature or seeking shade, allow them to tolerate variation in their external environment and maintain optimum metabolic rate.

Regulators

Regulators are organisms that can maintain their internal environment regardless of the external environment.
Regulators use metabolism to control their internal environment, which increases the range of possible ecological niches they can occupy.

Regulators require energy to achieve homeostasis and as a result have high metabolic costs. The control mechanism by which homeostasis is achieved is called negative feedback.

Negative feedback control

Negative feedback systems have monitoring centres with receptor cells, a system for sending messages and effectors. The receptors are special cells that constantly monitor the internal environment and detect changes. If a change from the optimum is detected, a corrective mechanism is switched on and messages are sent to the effectors. The messages can either be hormones in the blood or nerve impulses. The effectors are the parts of the body, such as muscles or glands, that respond to the messages. The effectors respond to correct the change and return conditions to their optimum. Figure 2.26 shows the general mechanism of negative feedback control.

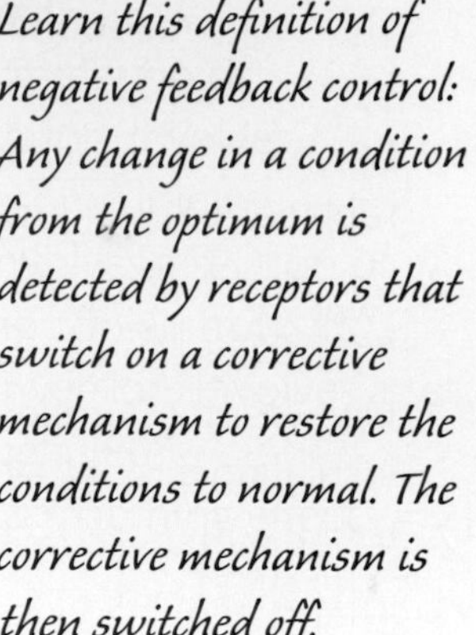

Learn this definition of negative feedback control: Any change in a condition from the optimum is detected by receptors that switch on a corrective mechanism to restore the conditions to normal. The corrective mechanism is then switched off.

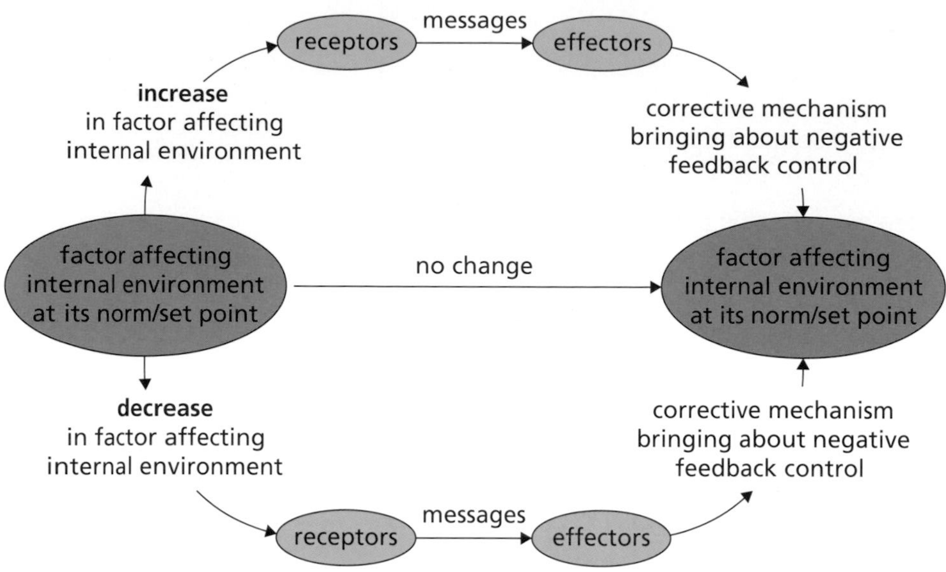

Figure 2.26 Negative feedback control

Thermoregulation in mammals

Mammals are thermoregulators and have a homeostatic mechanism to regulate their body temperature. Figure 2.27 shows the effect of external temperature on the body temperature of a mammal regulator compared with a reptile conformer.

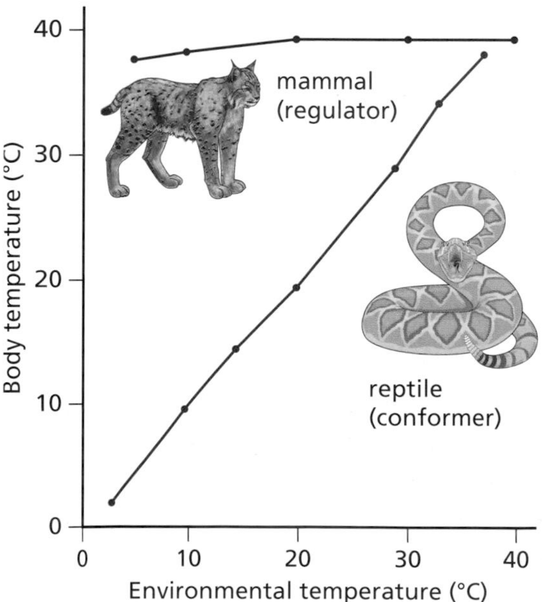

Figure 2.27 Temperature in a regulator and a conformer

The hypothalamus is the temperature-monitoring centre of the mammalian brain and contains thermoreceptors, which detect changes in blood temperature. The hypothalamus sends out electrical impulses through nerves to the effectors, skin and body muscles, which bring about corrective responses to return temperature to normal, as shown in Figure 2.28.

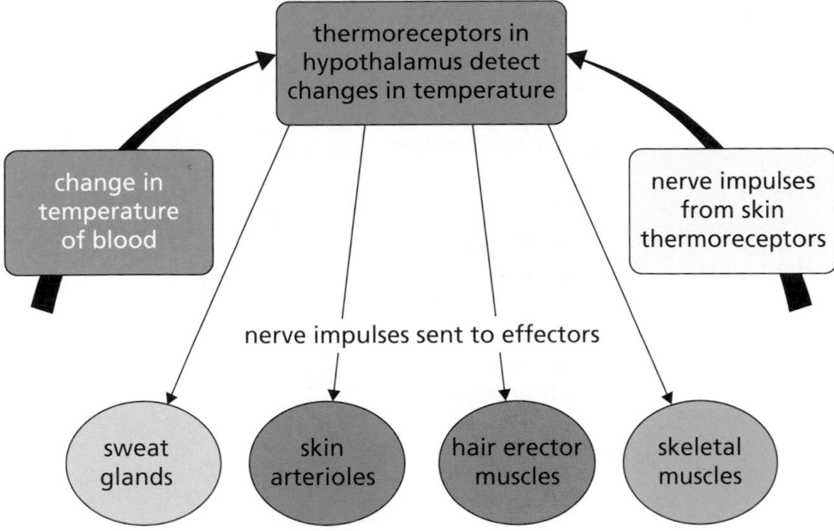

Figure 2.28 Role of hypothalamus

Corrective responses to an increase in body temperature

Hot conditions, exercise or illness can increase the body temperature above normal. The thermoreceptors in the hypothalamus detect this increase and send electrical impulses through nerves to the effectors. Sweat glands increase sweat production and heat from the body evaporates the water in the sweat, lowering the body temperature. Vasodilation occurs – the skin arterioles become dilated (wider) – allowing an increased volume of blood to flow through the capillaries on the skin surface, increasing heat loss. The metabolic rate can also be decreased so that less heat is produced. These corrective mechanisms lower the body temperature back to normal.

Corrective responses to a decrease in body temperature

If the body temperature falls below normal, the thermoreceptors in the hypothalamus detect the decrease and send electrical impulses through nerves to the effectors. The sweat glands decrease sweat production. Vasoconstriction occurs – the skin arterioles become constricted (narrower) – reducing the volume of blood that flows through the capillaries on the skin surface, so less heat is lost. Hair erector muscles contract to raise hairs, which trap air, providing insulation and reducing heat loss. Shivering of the skeletal muscles generates heat. There is also an increase in the metabolic rate, which increases heat production. These corrective mechanisms raise the body temperature back to normal. Figure 2.29 shows the homeostatic control of body temperature in humans.

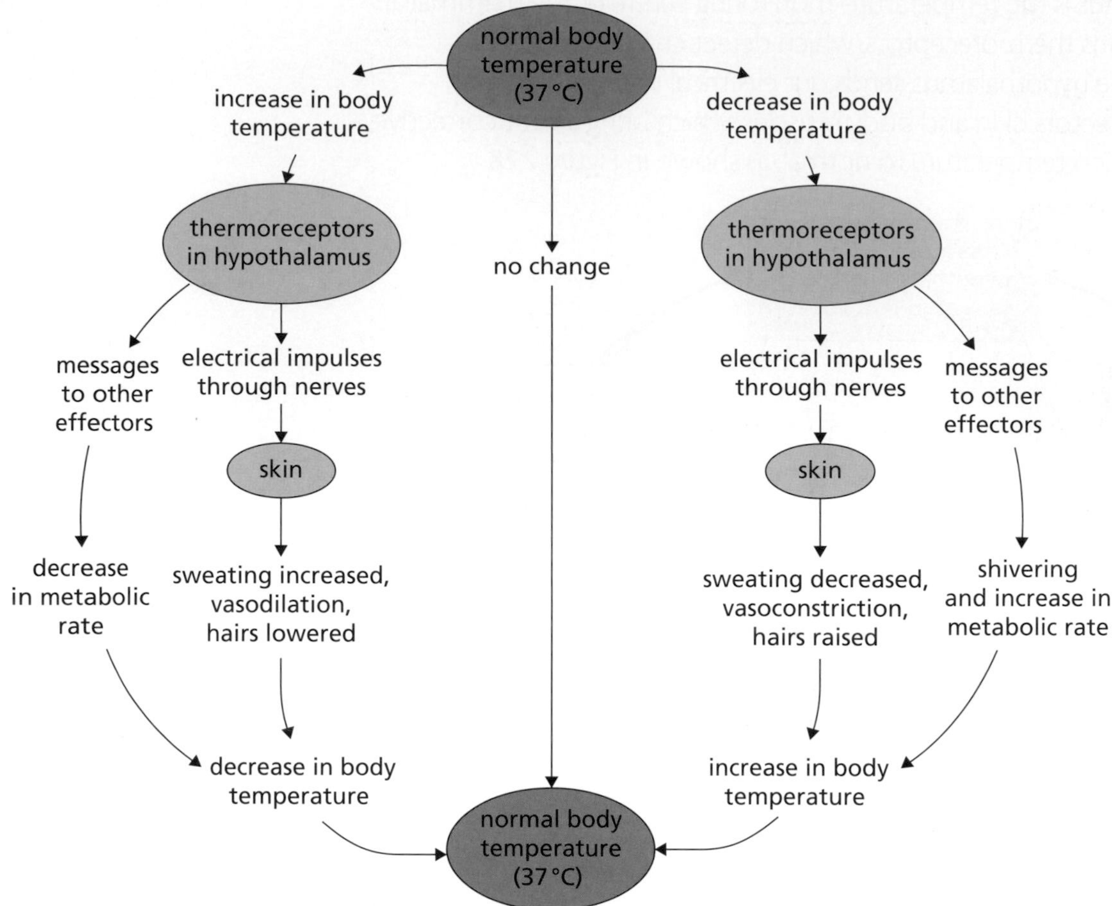

Figure 2.29 Thermoregulation in humans

Advantages of maintaining a constant body temperature

Enzymes have an optimum temperature at which they work best. Animals that can maintain the optimum temperature for enzyme activity can maintain a high metabolic rate.

Temperature also affects diffusion rates. Rates of diffusion of substances such as oxygen and carbon dioxide are faster at warmer temperatures and maintaining these contributes towards the ability to maintain high metabolic activity.

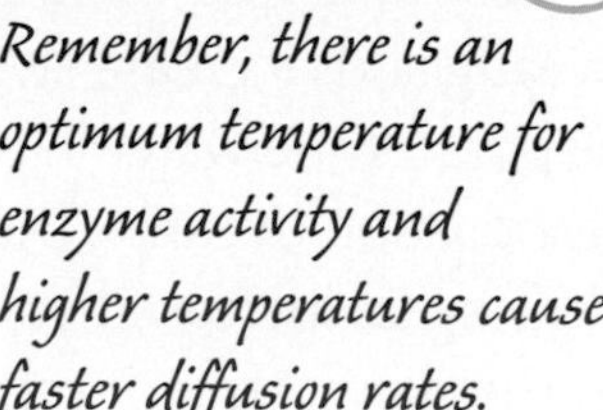

Remember, there is an optimum temperature for enzyme activity and higher temperatures cause faster diffusion rates.

Key words

Conformer – animal whose internal environment is dependent on its environment
Ecological niche – the way of life and the role of an organism in its community
Homeostasis – maintenance of a steady state or constant internal environment in the cells of a living organism
Hypothalamus – temperature-monitoring centre of the mammalian brain containing thermoreceptors, which detect changes in blood temperature
Negative feedback – control system for maintaining homeostasis in regulator organisms
Regulator – an animal that can control its internal environment and maintain homeostasis by using physiological mechanisms
Thermoreceptor – heat-sensitive cell in the hypothalamus of mammals
Thermoregulation – use of negative feedback in regulation of body temperature in mammals
Vasoconstriction – narrowing of the blood vessels resulting from contraction of the muscular wall of the vessels
Vasodilation – widening of blood vessels that results from relaxation of the muscular walls of the vessels

Questions 

Short answer (1 or 2 marks)

1 Mammals maintain their body temperature by homeostatic control.
  a) Name the temperature-monitoring centre in the brain that detects temperature changes. (1)
  b) In the following sentence, identify which of the alternatives in each pair makes the sentence correct.
  The corrective mechanism in response to a decrease in body temperature includes (vasoconstriction/vasodilation), which (increases/decreases) blood flow in the skin. (1)
  c) Describe how the messages from the temperature-monitoring centre are relayed to the effectors in the skin. (1)
  d) State **two** reasons why body temperature in humans is important to metabolic processes. (2)

2 a) Give **two** abiotic factors that affect the ability of an organism to maintain its metabolic rate. (2)
  b) Give the meaning of the term *homeostasis*. (1)
  c) Explain why conformers usually have a narrow ecological niche. (2)
  d) Give the name of the control system used by regulators to maintain a relatively stable internal environment. (1)

Longer answer (3–10 marks)

3 Give an account of the mechanisms of thermoregulation in mammals. (9)

4 Give an account of metabolism in conformers in relation to their ecological niches. (7)

Answers are on page 91.

Key Area 2.5

Metabolism and adverse conditions

Key points

1 Many environments vary beyond the tolerable limits for the normal metabolic activity of an organism. ☐
2 To cope with these cyclic or unpredictable fluctuations, organisms must have adaptations to survive or to avoid adverse conditions. ☐
3 To allow survival during a period when the costs of continued normal metabolic activity would be too high, the metabolic rate can be reduced. ☐
4 **Dormancy** is part of an organism's life cycle and is the stage associated with resisting or tolerating periods of environmental adversity. ☐
5 Dormancy can be **predictive** or **consequential**. ☐
6 Examples of dormancy include **hibernation** and **aestivation**. ☐
7 Hibernation is often defined in terms of mammals and is a common survival strategy in response to a metabolic energy crisis brought about by low temperatures and lack of food. ☐
8 Aestivation allows survival in periods of high temperature or drought. ☐
9 Daily **torpor** is a period of reduced activity in organisms with high metabolic rates. ☐
10 **Migration** avoids metabolic adversity by expending energy to relocate to a more suitable environment. ☐
11 Specialised techniques are used in studies of long-distance migration. For example, individual marking and types of tracking have been developed to overcome the difficulties involved in the study of migratory vertebrates and invertebrates. ☐
12 Experiments have been designed to investigate the **innate** and **learned** influences on migratory behaviour. ☐

Summary notes

Maintaining metabolism during environmental change

Many environments vary beyond the tolerable limits for normal metabolic activity of an organism. In some environments, extremes of temperature can result in lack of food and drought. To cope with these fluctuations, which can be cyclic or unpredictable, organisms have adaptations to survive them or avoid them.

1 Surviving adverse conditions

Dormancy

To allow survival during a period when the energy costs required for normal metabolic activity would be too high, the metabolic rate can be reduced. This can be achieved by a period of dormancy. During dormancy there is a decrease in metabolic rate, heart rate, breathing rate and body temperature. Dormancy is part of an organism's life cycle and can be predictive or consequential.

Predictive dormancy occurs when an organism becomes dormant before the onset of the adverse conditions. It occurs in advance of the adverse conditions and is usually genetically programmed. It is typical in predictable seasonal environments where the temperature and photoperiod can be used as environmental cues.

Consequential dormancy is when an organism becomes dormant after the onset of the adverse conditions. It is a typical response of organisms living in unpredictable environments.

Examples of dormancy in animals include hibernation, aestivation and torpor.

Hibernation is often defined in terms of mammals. It is a widespread and common survival strategy in response to the threat of a metabolic energy crisis brought about by winter, low temperatures and lack of food. Before hibernating, a mammal eats extra food and stores it as fat. During hibernation, metabolic rate is reduced, resulting in a decrease in body temperature, heart rate and breathing rate. This reduces energy expenditure and allows a mammal to survive the winter period.

Aestivation is a form of dormancy that allows some animals to survive in periods of high temperature or drought in the summer. It occurs not just because of food supply issues, but also because the conditions become too hot and dry for the animal to survive. The process typically involves burrowing into the ground, where the temperature stays cool, and reducing metabolic activity in a similar manner to hibernation.

Examples 

In Australia, the water-holding frog has an aestivation cycle to conserve energy. It buries itself in sandy ground in a secreted, water-tight, mucus cocoon during periods of hot, dry weather and remains in this state of dormancy until the arrival of the rain when conditions improve.

Lungfish are capable of a form of aestivation that allows them to live without water for as long as three years. Lungfish are fish that have lungs as well as gills, allowing them to breathe air. When its lake dries up, the fish burrows into the mud, secreting mucus until its entire body is covered. The mucus dries into a sack that holds moisture in. Even when the mud dries completely, the lungfish stays moist and breathes through a mucus tube.

Daily torpor is a period of reduced activity in organisms with high metabolic rates such as small birds and mammals. Daily torpor is similar to short-term hibernation.

Torpor results in a decrease in body temperature, heart rate and breathing rate and increases an organism's chances of survival by reducing the energy required to maintain a high metabolic rate. Small mammals and birds can reduce their rate of energy consumption during daily torpor by up to 90 per cent.

Example 

Hummingbirds have an extremely high metabolic rate, with a heart rate that can exceed 1200 beats per minute. Their energy consumption is so great that hummingbirds use daily torpor to conserve energy, even in the tropics.

2 Avoiding adverse conditions

Migration is a relatively long-distance movement of individuals, which usually takes place on a seasonal basis. Migration enables animals to avoid metabolic adversity brought about by lack of food and low temperatures by expending energy to relocate to a more suitable environment.

Hints & tips

Make sure you know the difference between ***surviving*** *adverse conditions and* ***avoiding*** *them.*

Innate and learned influences on migratory behaviour

Migration is found in all major animal groups, including birds, mammals, reptiles, amphibians, fish and insects. An organism's migratory behaviour is thought to be inherited or innate. Migratory behaviour is also influenced by learning, which is gained by experience.

Migration occurs in response to an external trigger stimulus such as the day length, local climate, the availability of food, the season of the year or for mating reasons. To be counted as a true migration, the movement of the animals should be an annual or seasonal occurrence, such as birds migrating for the winter or the annual migration of wildebeest for grazing. It can also involve a major habitat change as part of the life cycle, as in the case of the Atlantic salmon, which migrate from fresh water to seawater in an annual cycle.

Tracking migration

Scientists have developed specialised techniques to overcome the difficulties involved in the study of the long-distance migration of vertebrates and invertebrates. These include satellite tracking and the individual marking of animals, such as the ringing of a bird's leg with an identification tag, and attempting recaptures. Transmitters, attached to animals' bodies, which send out a signal that can be picked up by a receiver have also been developed. One advantage of these is that the animals being tracked do not need to be recaptured.

Key words

Aestivation – type of dormancy that allows animals to survive in periods of high temperature or drought

Consequential dormancy – dormancy that occurs in response to the onset of adverse conditions

Dormancy – reduction in metabolic rate made by organisms to tolerate adverse conditions (e.g. hibernation, aestivation)

Hibernation – response of an animal to survive adverse conditions by reduction of metabolic rate, brought about by low temperatures and lack of food

Innate – unlearned instinctive behaviour

Learned – behaviour of an individual organism not common to all members of its species and which is acquired by experience

Migration – response by an organism to avoid adverse conditions by relocating

Predictive dormancy – dormancy that occurs before the onset of adverse conditions triggered by environmental cues such as temperature and photoperiod

Torpor – period of reduced activity in organisms with high metabolic rates such as small birds and mammals

Questions ?

Short answer (1 or 2 marks)

1 **a)** Give the meaning of the term *dormancy*. (1)

b) Give **two** examples of dormancy. (2)

c) Describe the difference between predictive and consequential dormancy. (2)

d) Explain the benefit to some animals of being able to undergo daily torpor. (2)

2 **a)** Give the meanings of innate and learned behaviour in bird migration. (2)

b) Give **two** methods of tracking migratory animals. (2)

Longer answer (3–10 marks)

3 Give an account of the adaptations of organisms to surviving and avoiding adverse environmental conditions. (8)

4 Write notes on migratory behaviour with an example of how such behaviour can be tracked. (4)

Answers are on page 92.

Key Area 2.6

Environmental control of metabolism

Key points

1. Microorganisms are **Archaea**, bacteria and some species of eukaryotes. ☐
2. Microorganisms use a wide variety of substrates for metabolism and produce a range of products from their metabolic pathways. ☐
3. Microorganisms are useful because of their adaptability, ease of cultivation and speed of growth. ☐
4. The growth of microorganisms is influenced by the composition of their **growth medium** and by environmental conditions. ☐
5. Many microorganisms produce all the complex molecules required for **biosynthesis**, for example amino acids, vitamins and fatty acids. ☐
6. Other microorganisms require these to be supplied in the growth media. ☐
7. When culturing microorganisms, their growth media requires raw materials for biosynthesis as well as an energy source. ☐
8. An energy source is derived either from chemical substrates or from light in photosynthetic microorganisms. ☐
9. Culture conditions include **sterility**, control of temperature, oxygen levels and pH. ☐
10. Sterile conditions in **fermenters** reduce competition with desired microorganisms for nutrients and reduce the risk of spoilage of the product. ☐
11. When microorganisms are cultured in a liquid medium, they use up the substrate and nutrients available and release **metabolites** back into the medium. ☐
12. These changes result in the four characteristic phases of growth: the **lag**, **log** (**exponential**), **stationary** and **death** phases. ☐
13. During the lag phase enzymes are induced to metabolise the available substrates. ☐
14. The log (exponential) phase shows the most rapid growth of microorganisms due to plentiful nutrients. ☐
15. The stationary phase occurs due to the nutrients in the culture media becoming depleted and the production of toxic metabolites. ☐
16. **Secondary metabolites** are also produced, such as antibiotics. In the wild these metabolites confer an ecological advantage by allowing the microorganisms which produce them to out-compete other microorganisms. ☐
17. The death phase occurs due to the toxic accumulation of metabolites or the lack of nutrients in the culture. ☐
18. **Semi-logarithmic graphs** are used in producing or interpreting growth curves of microorganisms. ☐

⇨

19 To estimate microbial population number, cell counts are made. ☐

20 **Viable cell counts** involve counting only the living microorganisms, whereas **total cell counts** involve counting viable and dead cells. ☐

21 Only viable cell counts show a death phase when cell numbers are decreasing. ☐

Summary notes

Metabolism in microorganisms

Microorganisms include Archaea, bacteria and some species of eukaryote such as yeasts and protozoans. Microorganisms include species that use a wide range of substrates for metabolism and produce a wide range of products from their metabolic pathways.

As a result of their adaptability, microorganisms are found in a wide range of ecological niches and can be used for a variety of research and industrial applications because they reproduce and grow quickly. They are easy to culture and they produce many different useful products.

Environmental control of metabolism

Culture media

The growth of microorganisms is influenced by the composition of their growth medium and the environmental conditions in which they are being cultured. Microorganisms require an energy source and a supply of raw materials in their growth medium. They use the raw materials for the biosynthesis of more complex substances such as proteins and nucleic acids.

Many microorganisms can produce all the complex molecules required for biosynthesis, including the amino acids required for protein synthesis, vitamins and fatty acids, from simple chemical compounds in growth media. Other microorganisms require specific complex compounds such as vitamins or fatty acids to be supplied in the growth media.

An energy source is derived either from chemical substrates such as carbohydrates or from light in the case of photosynthetic microorganisms.

Culture conditions

Culture conditions include sterility to eliminate any effects of contaminating microorganisms, and the control of temperature, oxygen levels by aeration and pH by buffers or the addition of acid or alkali. Sterile conditions in fermenters reduce competition with desired microorganisms for nutrients and reduce the risk of spoilage of the product.

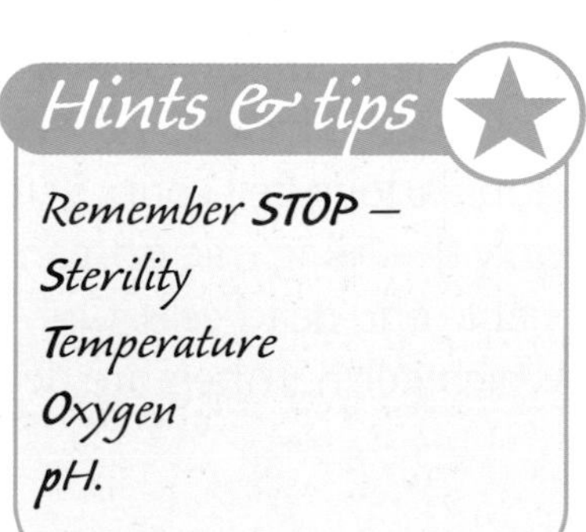

Industrial fermenters or bioreactors, which are used to culture microorganisms on a huge scale, are controlled automatically by computers. Sensors monitor the culture conditions and maintain the factors affecting growth at their optimum level, as shown in Figure 2.30.

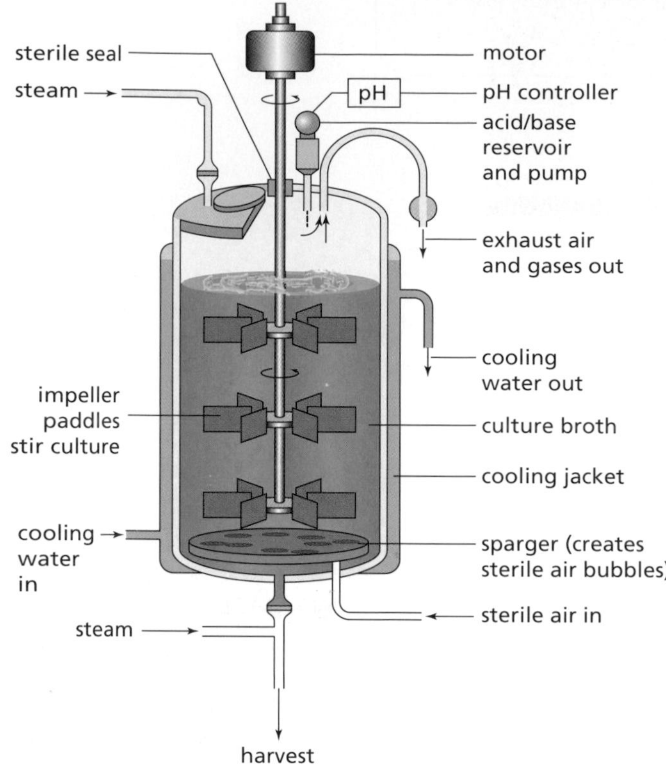

Figure 2.30 Industrial fermenter

Growth

Growth can be defined as the irreversible increase in dry biomass of an organism.

Growth of unicellular organisms such as bacteria and yeast is recorded by measuring the increase in cell number in a given period of time.

> **Hints & tips**
>
> *The dry mass is a more reliable measurement of growth than the fresh mass because fresh mass includes the water content in the cells of the organisms, which can vary independently of growth.*

Phases of microorganism growth

The time it takes for a unicellular organism to divide into two is called the doubling or **generation time**. When microorganisms are cultured in a liquid medium, they use up the substrate and nutrients available and release metabolites back into the medium. These changes result in the four characteristic phases of growth: the lag, log (exponential), stationary and death phases, as shown in Figure 2.31. To estimate microbial populations, cell counts are made. Viable cell counts involve counting only the living microorganisms, whereas total cell counts involve counting viable and dead cells. Only a viable cell count can show a death phase in which cell numbers are decreasing.

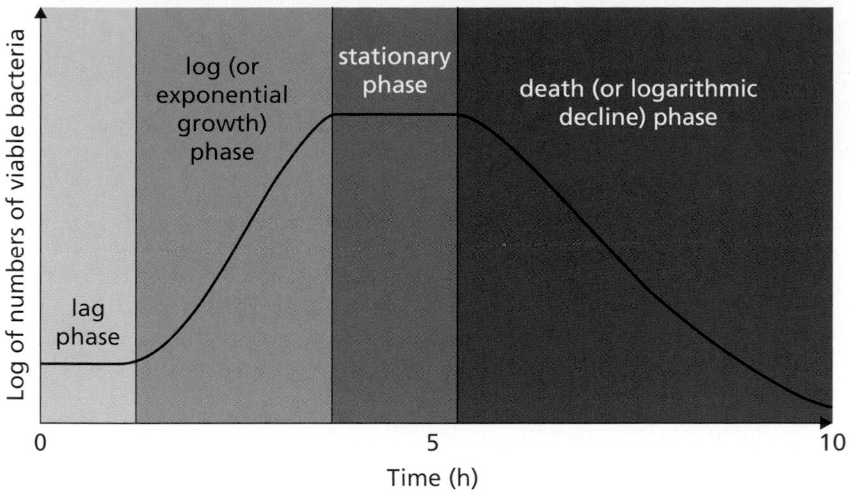

Figure 2.31 Phases of growth in microorganisms

In the lag phase of growth the microorganisms adjust to the conditions of the culture by inducing the production of enzymes that metabolise the available substrates. No cell division occurs at this stage.

In the log or exponential phase of growth the population doubles with each round of cell division. The log (exponential) phase contains the most rapid growth of microorganisms due to plentiful nutrients. The rate depends on the culture medium used and the temperature.

In the stationary phase the culture medium becomes depleted and nutrients or oxygen start to run out. Metabolites released by the microorganisms begin to accumulate and these may be toxic to the microorganism. Secondary metabolites are also produced, such as antibiotics. In the wild these metabolites confer an ecological and selective advantage by allowing the microorganisms which produce them to out-compete other microorganisms.

The stationary phase is reached when the rate of production of new cells is equal to the death rate of the older cells and there is therefore no increase in cell number in the culture.

In the death phase the lack of substrate and the toxic accumulation of metabolites causes the death rate of the cells to be greater than the production of any new cells. More cells die than are being produced. By the end of this phase a bacterial population might be completely eliminated or resistant spores might remain.

Figure 2.32 shows the exponential growth phase of a unicellular culture using semi-log graph paper, which gives a straight-line graph. Note that on the cell number scale the division between 1 and 10 is the same size as that between 10 and 100.

Hints & tips

Remember to mention that enzymes are synthesised in the log phase.

Hints & tips

*Remember **SS** = **S**econdary metabolites produced in the **S**econdary phase*

Key links

There is more about selective advantage in Key Area 1.7 (page 26).

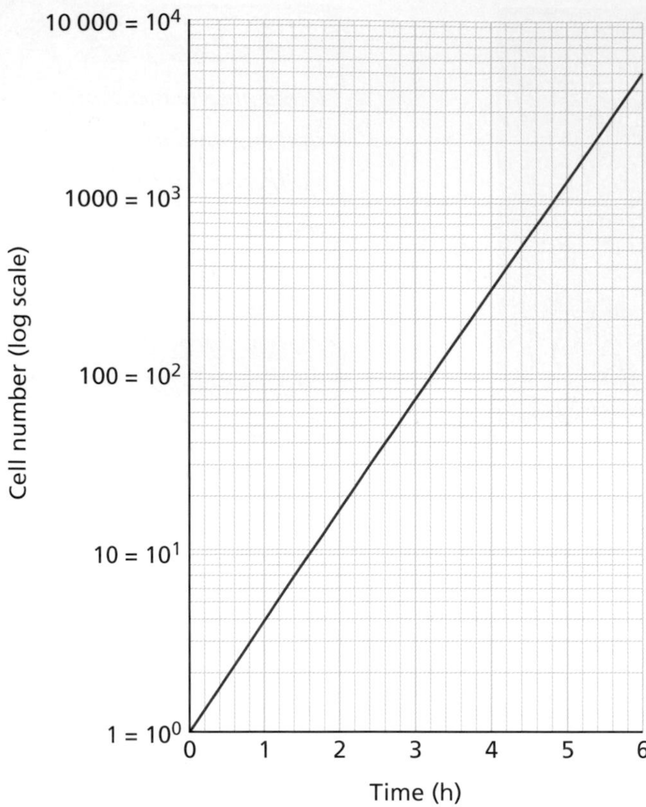

Figure 2.32 Exponential growth plotted on semi-log graph paper

Primary and secondary metabolism

Primary metabolism refers to the metabolism of a microorganism that occurs during the lag and log phases of its growth. This is when it breaks down the available substrate to obtain energy and produces primary metabolites that can be used for the biosynthesis of substances such as proteins and nucleic acids.

Secondary metabolism occurs at the end of the log phase and during the stationary phase of growth. Secondary metabolism produces substances that are not associated with growth but which may confer an ecological advantage. Examples of these secondary metabolites include antibiotics, which inhibit the growth of other species of bacteria and so reduce competition for the available resources.

Key words

Archaea – a domain of single-celled microorganisms
Biosynthesis – a multi-step, enzyme-catalysed process in living organisms in which substrates are converted into more complex products
Death phase – phase of microorganism growth in which death rate of cells exceeds rate of cell division
Exponential growth – growth phase of microorganisms involving a rapid geometric increase in numbers
Fermenters – vessels for growing large quantities of microorganisms under optimum conditions
Generation time – time taken for a microorganism cell to divide
Growth medium – substance which provides microorganisms with an energy source and raw materials for biosynthesis
Lag phase – phase when microorganisms adjust to the conditions of the culture by inducing enzymes that metabolise the available substrates
Log phase – exponential phase of microorganism growth
Metabolite – a substance produced by metabolism or a substance necessary for a particular metabolic process
Secondary metabolite – substance, not associated with growth, produced during the stationary phase of growth of a culture of microorganisms, for example antibiotics
Semi-logarithmic graph – graph with one logarithmic scale and one linear scale. A logarithmic scale is a non-linear scale which is used when there is a large range of quantities
Stationary phase – phase of microorganism growth during which secondary substances can be made
Sterility – not containing contaminating microorganisms
Total cell count – total number of cells in a culture including viable (live) cells and dead cells
Viable cell count – number of live cells from a total cell count

Questions ?

Short answer (1 or 2 marks)

1 a) Describe **two** features of microorganisms that make them useful for a variety of research and industrial uses. (2)
b) Give **two** complex compounds that are sometimes added to culture media to enable certain microorganisms to grow. (2)
c) Explain why sterile conditions must be maintained in the culture of a microorganism. (1)
d) Explain why temperature, pH and oxygen levels must be monitored and controlled during the culture of microorganisms. (2)

2 a) Explain why the number of microorganisms remains constant during the lag phase of microorganisms grown in culture. (2)
b) Explain why substances produced during secondary metabolism might give an ecological advantage to a microorganism. (2)
c) Name the type of cell count that shows the death phase of a population of microorganisms in a culture vessel. (1)

Longer answer (3–10 marks)

3 Give an account of the different culture conditions required for the growth of microorganisms. (9)
4 Give an account of the phases of growth of microorganisms cultured in a fermenter. (7)

Answers are on page 93.

Key Area 2.7

Genetic control of metabolism

Key points

1 Wild strains of microorganisms can be made more useful by **mutagenesis** or **recombinant DNA technology**. ☐

2 Exposure to UV light, other forms of radiation or mutagenic chemicals results in mutations, some of which may produce an improved strain of microorganism. ☐

3 Recombinant DNA technology involves the use of recombinant plasmids and **artificial chromosomes** as **genetic vectors**. ☐

4 A genetic vector is a DNA molecule that is used to carry foreign genetic information into another cell. Both plasmids and artificial chromosomes are used as vectors during recombinant DNA technology. ☐

5 Artificial chromosomes are preferable to plasmids as vectors when larger fragments of foreign DNA are required to be inserted. ☐

6 **Restriction endonucleases** cut open plasmids and specific genes out of chromosomes, leaving sticky ends.

7 Complementary sticky ends are produced when the same restriction endonuclease is used to cut open the plasmid and the gene from the chromosome. ☐

8 The enzyme ligase seals the gene into the plasmid.

9 Recombinant plasmids and artificial chromosomes contain **restriction sites**, **regulatory sequences**, an **origin of replication** and **selectable markers**. ☐

10 Restriction sites contain target sequences of DNA where specific restriction endonucleases cut. ☐

11 Regulatory sequences control gene expression and an origin of replication allows self-replication of the plasmid or artificial chromosome. ☐

12 Selectable markers, such as antibiotic resistance genes, protect the microorganism from a selective agent (for example, an antibiotic) that would normally prevent growth or kill it. ☐

13 Selectable marker genes present in the vector ensure only microorganisms that have taken up the vector grow in the presence of the selective agent. ☐

14 As a safety mechanism, **safety genes** are often introduced that prevent the survival of the microorganism in an external environment. ☐

15 Plant or animal recombinant DNA in bacteria can result in polypeptides that are folded incorrectly and are non-functional. ☐

16 Recombinant yeast cells can be used to produce active (functional) forms of the protein. ☐

Summary notes

Genetic control of metabolism

Wild strains of microorganisms, with the potential to be used in industry in the production of a desirable product, can be improved by processes such as mutagenesis, or recombinant DNA technology.

Mutagenesis

Mutagenesis is the process of inducing mutations. Exposure to ultraviolet (UV) light, other forms of radiation or mutagenic chemicals results in random mutations, some of which may produce an improved strain of microorganism with desirable qualities, which can then be selected and cultured for use. Improved microorganisms for an industrial application could refer to their ability to be cultured in a low-cost medium or to the fact that the production of the required product is increased.

Recombinant DNA technology

Microorganisms can be transformed by transferring plant or animal gene sequences to them. Recombinant DNA technology involves the joining together of DNA molecules from two different species. The DNA sequences used in the construction of recombinant DNA molecules can originate from any species. It can then be inserted into a host organism to produce new genetic combinations that are of value to science, medicine, agriculture and industry. Recombinant DNA technology involves the use of recombinant plasmids and artificial chromosomes as vectors. A vector is a DNA molecule that is used to carry foreign genetic information into another cell. Artificial chromosomes are preferable to plasmids as vectors when larger fragments of foreign DNA are required to be inserted.

Plant or animal gene sequences can be transferred to microorganisms to produce plant or animal proteins.

Control of gene expression in recombinant plasmids and artificial chromosomes

Extra-chromosomal DNA molecules can be transferred to microorganisms.

Genetic engineers can identify, locate and extract a gene coding for a desirable characteristic and then insert and seal it into the DNA of a vector, such as a bacterial plasmid or an artificial chromosome, before inserting it into the host cell such as a bacterium. Figure 2.33 shows some of the stages involved in recombinant DNA technology. Artificial chromosomes are preferable to plasmids as vectors when larger fragments of foreign DNA are required to be inserted.

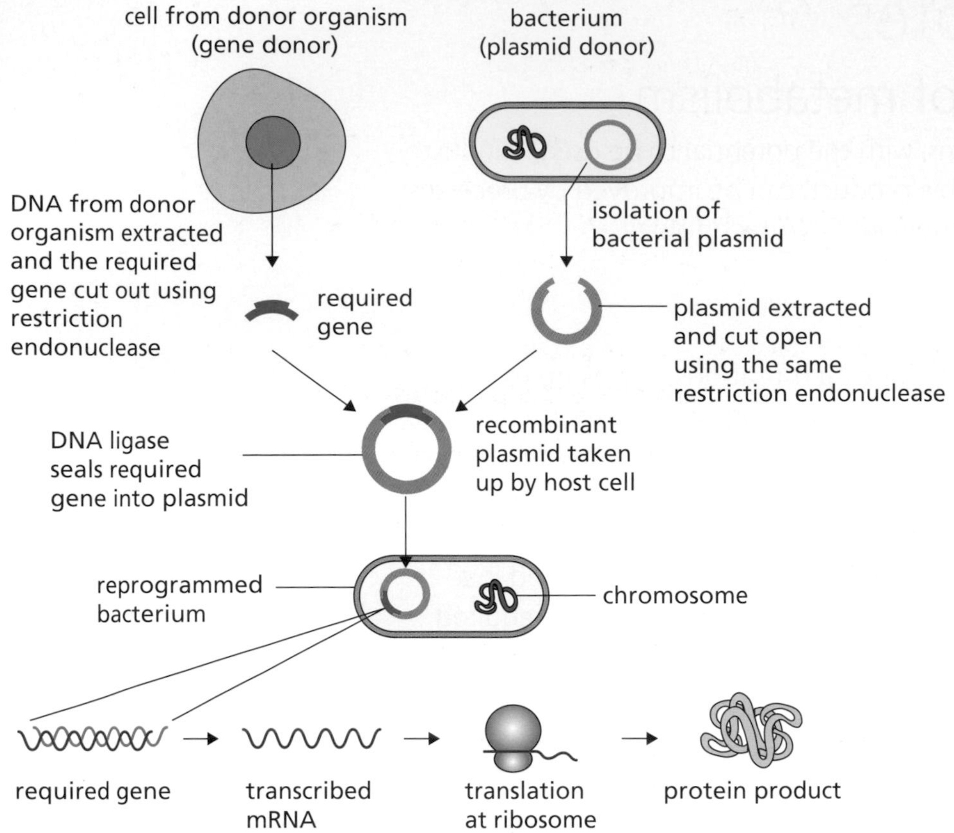

Figure 2.33 Stages in recombinant DNA technology

The recombinant plasmid is called a vector because it carries the DNA from the donor organism into the host cell.

To be an effective vector, the plasmid must contain restriction sites and selectable marker genes, in addition to genes for self-replication and regulatory sequences, to allow the control of gene expression.

Figure 2.34 shows the features of an effective plasmid vector.

The plasmid must have a restriction site, which is a location on the plasmid that can be cut open by the same restriction endonuclease used to extract the gene from the donor's DNA. Restriction endonucleases cut target sequences of DNA, leaving sticky ends. Treatment of vectors with the same restriction endonuclease forms complementary sticky ends that are then combined with the target sequences using DNA ligase to form recombinant DNA.

Figure 2.34 Features of an effective plasmid vector

The plasmid acting as a vector also requires a selectable marker gene. A selectable marker gene is a gene which is deliberately transferred along with the required gene during the process of genetic engineering. It is easily recognised and used to identify those cells to which the gene has been successfully transferred. Thus, a selectable marker gene is a gene used to determine if a DNA sequence has been successfully inserted into the host organism's DNA.

There are two types of selectable marker genes.

Selectable markers such as antibiotic resistance genes protect the microorganism from a selective agent (antibiotic) that would normally kill it or prevent its growth. Only one in several million cells may take up donor DNA and, rather than checking every single cell, the genetic engineers use a selective agent to kill all cells that do not contain the new DNA, which means that only the modified ones survive. Antibiotics are the most common selective agents.

A selectable marker gene for screening is a gene that makes the cell containing the gene look different, for example a marker gene that produces a green fluorescent protein which makes the modified cells glow green under UV light.

The plasmid also requires an origin of replication. This contains genes for the self-replication of the plasmid, which is essential for the future copying of the plasmid. As a safety mechanism, genes are often introduced that prevent the survival of the microorganism in an external environment.

Plant or animal recombinant DNA expressed in bacteria can result in the production of polypeptides that are folded incorrectly and are non-functional. These proteins can be produced more successfully in recombinant yeast cells, which are eukaryotic and so have the enzymes needed to produce active (functional) forms of the protein.

Example 

Ethical issues

Ethics is about the application of moral frameworks concerning the principles of conduct governing individuals and groups. In the context of gene technology, it is to do with issues of whether it is right or wrong to conduct research into and develop these technologies.

The following table shows some of the ethical arguments that can be linked to genetic modification technology.

Arguments in favour of the technology	Arguments against the technology
It might improve nutrition and food security by increasing quantity and quality of food	The potential impact of the technology is unknown and many aspects of it, such as the safety of foods or drugs, remain to be understood
It might improve the environment by allowing reduction of the use of pesticides or fertilisers	The risks of the organisms or the genes they contain escaping are too great and could not be reversed
It might improve health by the production of drugs that are otherwise difficult to produce	Genes are self-perpetuating, and the risks that they might bring in the future are unknown

Key words

Artificial chromosome – used as a vector to carry foreign genetic information into another cell
Genetic vector – a DNA molecule such as a plasmid or artificial chromosome used to carry foreign genetic information into another cell
Mutagenesis – the process of inducing mutations by exposure to mutagenic agents such as ultraviolet (UV) light, other radiation, or certain chemicals
Origin of replication – site which allows self-replication of a plasmid or artificial chromosome
Recombinant DNA technology – involves the joining together of DNA molecules from two different species
Regulatory sequences – DNA sequences which control gene expression
Restriction endonuclease – enzyme which cuts target sequences of DNA from a chromosome or is used to open a plasmid
Restriction sites – target sequences of DNA where specific restriction endonucleases cut
Safety genes – genes which are introduced that prevent the survival of the microorganism in an external environment
Selectable markers – genes present in the vector that ensure only microorganisms which have taken up the vector grow in the presence of the selective agent, for example an antibiotic

Questions ?

Short answer (1 or 2 marks)

1 **a)** State **two** methods by which wild strains of microorganisms could be improved. (2)
b) Explain what is meant by the term *mutagenesis*. (1)
c) Give **one** example of a mutagenic agent that can increase the rate of mutation in an organism. (1)

2 Bacterial plasmids can be modified by inserting a gene from another organism.
a) Two enzymes are required to produce the modified plasmid.
(i) Name the enzyme that cuts the plasmid at specific restriction sites. (1)
(ii) Name the enzyme that seals the gene from the donor organism into the plasmid. (1)
b) Give **two** reasons for the transfer of gene sequences to microorganisms **other than** for the production of plant or animal proteins. (2)
c) **(i)** Explain why animal DNA that has been transferred to bacteria might produce proteins that are not functional. (1)
(ii) Suggest how this problem might be overcome. (1)

Longer answer (3–10 marks)

3 Give an account of the production of protein by recombinant DNA technology. (9)

Answers are on page 94.

Answers

Key Area 2.1a

Short answer questions

1 a) sequence of chemical reactions controlled by enzymes [1]

b) anabolic pathways are biosynthetic processes/involve the building up of complex molecules from simpler substances *and* catabolic pathways involve the breakdown of complex molecules into simpler substances [1]

anabolic pathways require the input of energy *and* catabolic pathways usually release energy [1]

2 a) pores; pumps; enzymes [any 2 = 2]

b) mitochondria; chloroplasts; nucleus [any 2 = 2]

Longer answer questions

3 a) a metabolic pathway is a sequence of chemical reactions controlled by enzymes; chemical reactions are anabolic and catabolic; anabolic pathways are biosynthetic processes/involve the building up of complex molecules from simpler molecules; anabolic pathways require the input of energy; catabolic pathways involve the breakdown of complex molecules into simpler molecules; catabolic pathways usually release energy [any 4 = 4]

b) composed of phospholipid; and protein; protein can be enzymes; can be pumps; can be pores [any 4 = 4]

[total = 8]

Answers

Key Area 2.1b

Short answer questions

1 each step in a metabolic pathway is controlled by a specific enzyme; each enzyme is coded for by a gene; order of bases in the gene determines the order of amino acids, which determines structure, shape and function of the protein/enzyme [any 2 = 2]

2 a) enzyme is flexible and so the active site can change shape; substrate induces the active site to change shape; active site can alter the position or orientate the substrate molecules so that they fit more closely [any 2 = 2]

b) increase in substrate concentration drives the chemical reaction in the direction of the end product; increases the rate of reaction [1 each = 2]

c) active site can alter the position of/ orientate the substrate molecules so that they fit more closely; activation energy is lowered when an enzyme is involved [1 each = 2]

Longer answer questions

3 as the concentration of the inhibitor increases, the rate of reaction decreases; because fewer active site are available for reaction; at high inhibitor concentration the reaction stops; because all active sites are blocked/no longer fit substrate [4]

4 enzyme activity depends on the flexible/ dynamic shape of enzyme molecules; substrate has an affinity for the active site; induced fit; active site orientates the reactants; enzymes lower the activation energy; products have a low affinity for the active site; substrate and product concentration affects the direction and rate of reactions *or* increasing the substrate concentration increases/speeds up/drives forward the rate of the reaction; enzymes act in groups/ multi-enzyme complex [any 6 = 6]

in competitive inhibition the inhibitor resembles the substrate molecule; inhibition is reduced by increase in substrate concentration; in non-competitive inhibition the shape of the active site is changed; product inhibition/feedback inhibition [any 3 = 3]

[total = 9]

Answers

Key Area 2.2

Short answer questions

1 a) phosphorylation of intermediates in glycolysis uses 2 ATP; later reactions in glycolysis result in the direct regeneration of 4 ATP for every glucose molecule and so this gives a net gain of 2 ATP [1 each = 2]

b) dehydrogenase enzymes remove hydrogen ions from a substrate along with electrons [1]

c) NAD transports H^+ and electrons; to the electron transport chain [1 each = 2]

2 a) ATP synthase [1]

b) electrons pass down the chain of electron acceptors; releasing their energy, which is then used to pump hydrogen ions (H^+) across the inner mitochondrial membrane [1 each = 2]

c) oxygen acts as the final electron acceptor and combines with hydrogen ions and electrons to form water [1]

Longer answer questions

3 glycolysis is the breakdown of glucose to pyruvate; 2 ATP molecules are used to phosphorylate intermediates in glycolysis; an energy investment phase; 4 ATP molecules are produced/generated/made in a pay-off stage; H carried away by NAD [any 3 = 3]

if oxygen is available/in aerobic conditions pyruvate progresses to the citric acid cycle; pyruvate is converted/broken down to an acetyl group; acetyl group combines with coenzyme A; acetyl (coenzyme A) combines with oxaloacetate to form citrate; citric acid cycle is enzyme controlled/involves dehydrogenases; ATP generated/synthesised/produced/released at substrate level in the citric acid cycle; carbon dioxide is released from the citric acid cycle; oxaloacetate is regenerated; NAD/NADH transports electrons/transports hydrogen ions to the electron transport chain [any 6 = 6]

[total = 9]

4 electron transport chain on the inner membrane of the mitochondria; electron transport chain is a collection of proteins attached to a membrane; NADH releases electrons to the electron transport chain on the inner mitochondrial membrane; electrons pass down the chain of electron acceptors, releasing energy; energy is used to pump hydrogen ions (H^+) across the inner mitochondrial membrane; return flow of the hydrogen ions (H^+) back into the matrix drives the enzyme ATP synthase; synthesis of ATP from ADP + Pi; this stage produces most of the ATP generated by cellular respiration; final electron acceptor is oxygen; oxygen combines with hydrogen ions and electrons to form water [any 7 = 7]

regeneration of ATP from ADP + Pi uses the energy released from cellular respiration; ATP is used to transfer the energy from cellular respiration to synthetic pathways/cellular processes where energy is required; breakdown of ATP to ADP and phosphate/Pi releases energy [any 2 = 2]

[total = 9]

Answers

Key Area 2.3

Short answer questions

1 a) metabolic rate of an organism is the amount of energy consumed in a given period of time [1]

b) can be measured in terms of the oxygen consumed in a given period of time; carbon dioxide produced in a given period of time; the energy released as heat in a given period of time [any 2 = 2]

2 a) respirometer measures rate of respiration by the volume of oxygen consumed per unit time [1]

b) calorimeter measures metabolic rate directly by the increase in an animal's body temperature per unit time [1]

Longer answer questions

3 Fish circulation and heart structure: single circulatory system/heart to gills to body to heart; heart with only two chambers/atrium and a ventricle; loss of pressure a problem

Amphibian circulation and heart structure: incomplete double circulatory system; heart with three chambers/right and left atrium and one ventricle; pressure maintained; tissue blood is incompletely oxygenated

Mammal circulation and heart structure: complete double circulatory system; heart has four chambers/two atria and two ventricles; pressure maintained; tissue blood completely oxygenated [any 9 = 9]

Answers

Key Area 2.4

Short answer questions

1 a) hypothalamus [1]

b) vasoconstriction and decreases [1]

c) electrical/nerve impulses [1]

d) enzymes controlling the metabolism are maintained at their optimum temperature; affects rates of diffusion [1 each = 2]

2 a) temperature; salinity; pH [any 2 = 2]

b) homeostasis is the maintenance of a steady state/a constant internal environment within an organism [1]

c) conformers' internal environments are directly dependent on their external environment; conformers lack the ability to tolerate change should it occur [1 each = 2]

d) negative feedback control [1]

Longer answer questions

3 temperature-monitoring centre/thermoreceptors are located in the hypothalamus *or* information about temperature detected/received by hypothalamus; mammals derive most of their body heat from respiration/metabolism/chemical reactions; nerve message/communication/impulse sent to skin/effectors; vasodilation/widening of blood vessels to skin in response to increased temperature; more/increased blood to skin/extremities; increased/more heat radiated from skin/extremities *or* vasoconstriction/narrowing of blood vessels to skin in response to decreased temperature; less blood to skin/extremities; decreased/less heat radiated from skin/extremities; increased temperature/body too hot leads to increase in sweat production *or* converse; increase in heat loss due to evaporation of water in sweat *or* converse; decrease in temperature causes hair erector muscles to raise/erect hair; traps warm air *or* forms insulating layer; decrease in temperature causes muscle contraction/shivering, which generates heat/raises body temperature; temperature regulation involves/is an example of negative feedback [any 9 = 9]

4 ability of an organism to maintain its metabolic rate is affected by external abiotic factors such as temperature/salinity/pH; conformer's internal environment is ⇨

dependent on its external environment; conformers cannot alter their metabolic rate using physiological means; conformers usually have a narrow ecological niche/ limited range; conformers lack the ability to tolerate change should it occur; conformers live in stable environments; conformers do not use energy-requiring physiological mechanisms to alter their metabolic rate; conformers have low metabolic energy costs; many conformers manage to maintain their optimum metabolic rate by employing certain behavioural responses [any 7 = 7]

Answers

Key Area 2.5

Short answer questions

1 a) dormancy is part of an organism's life cycle and is the stage associated with resisting/ tolerating periods of environmental adversity [1]

b) hibernation; aestivation; torpor [any 2 = 2]

c) predictive dormancy is when an organism becomes dormant before onset of the adverse conditions; consequential dormancy is when an organism becomes dormant after the onset of adverse conditions [1 each = 2]

d) torpor increases an organism's chances of survival by reducing the energy required to maintain a high metabolic rate (or metabolic rate can be reduced); when the conditions result in the cost of metabolic activity being too high [2]

2 a) innate means that it is inherited; migratory behaviour is also influenced by learning, which is gained by experience [1 each = 2]

b) ringing; tagging; transmitters [any 2 = 2]

Longer answer questions

3 Surviving adverse conditions: some environments vary beyond tolerable limits; the extremes of conditions do not allow the normal metabolism of the organisms present; variation in conditions can be cyclical or unpredictable; metabolic rate can be reduced when conditions would make the cost of metabolic activity too high; dormancy may be predictive or consequential; example from hibernation *or* aestivation; daily torpor is a period of reduced activity; example of an organism and the adverse conditions that it survives [any 6 = 6]

Avoiding adverse conditions: migration avoids metabolic adversity by relocation; methods used to study migration – marking example *or* tracking example; example of a vertebrate animal and the adverse conditions that it avoids [any 2 = 2]

[total = 8]

4 Some organisms avoid adverse conditions by undertaking migration; migration uses energy to relocate to a more suitable environment; migratory behaviour can be innate and learned; example of techniques to track migration – satellite tracking or leg rings (ringing and recapture) [1 each = 4]

Answers

Key Area 2.6

Short answer questions

1 a) use a wide range of substrates for metabolism; produce a wide range of products from their metabolic pathways; ease of cultivation; speed of growth; adaptability [any 2 = 2]

b) amino acids; vitamins; fatty acids [any 2 = 2]

c) to eliminate any contaminating/competing microorganisms [1]

d) optimum temperature and pH for microorganism enzymes; oxygen for aerobic respiration of microorganisms [1 each = 2]

2 a) no cell division takes place in the lag phase; enzymes are induced to metabolise the available substrates [2]

b) results in the production of secondary metabolites such as antibiotics; inhibit the growth of bacteria and so reduce competition for the available resources [1 each = 2]

c) viable cell count [1]

Longer answer questions

3 require an energy source; energy is derived from substrates such as carbohydrates/light in the case of photosynthetic microorganisms; supply of raw materials for the biosynthesis of proteins/nucleic acids; many microorganisms only require simple chemical compounds in growth media; other microorganisms require specific complex compounds; example of complex compound, for example vitamins/fatty acids; sterility to eliminate any effects of contaminating microorganisms; control of temperature/pH/oxygen; control of oxygen levels by aeration; control of pH by buffers or the addition of acid or alkali [any 9 = 9]

4 growth is recorded by measuring the increase in cell number in a given period of time; the time it takes for a unicellular organism to divide into two is called the doubling or generation time; the lag phase of growth is where the microorganisms induce the production of enzymes that metabolise the substrates; no cell division occurs at this stage; log or exponential phase of growth is where the population doubles with each cell division; stationary phase is where the culture medium becomes depleted/nutrients or oxygen start to run out; stationary phase is reached when the rate of production of new cells is equal to the death rate of the older cells; death phase occurs due to lack of substrate/toxic accumulation of metabolites; more cells die than are being produced [any 7 = 7]

Answers

Key Area 2.7

Short answer questions

1 a) mutagenesis; recombinant DNA technology [1 each = 2]

b) the process of inducing mutations [1]

c) exposure to ultraviolet (UV) light; other forms of radiation; mutagenic chemicals [any 1 = 1]

2 a) (i) restriction endonuclease [1]

(ii) (DNA) ligase [1]

b) genes preventing the survival of the microorganism in an external environment; selectable marker genes to ensure that only microorganisms which have taken up the vector or plasmid grow OR named example [1 each = 2]

c) (i) production of polypeptides that are folded incorrectly [1]

(ii) use yeast instead of bacteria [1]

Longer answer questions

3 recombinant DNA technology involves the joining together of DNA molecules from two different species; plant or animal gene sequences can be transferred to microorganisms to produce plant or animal proteins; selectable marker genes can be added to ensure that only microorganisms which have taken up the vector or plasmid grow OR named example; genes to prevent the survival of the microorganism in an external environment can be introduced as a safety mechanism; recombinant plasmids or artificial chromosomes act as vectors; vector carries the DNA from the donor organism into the host cell; vectors must contain restriction sites/ marker genes/genes for self-replication and regulatory sequences; restriction endonucleases cut target sequences of DNA, leaving sticky ends; treatment of vectors with the same restriction endonuclease; complementary sticky ends are then combined using DNA ligase to form recombinant DNA [any 9 = 9]

Practice course assessment: Area 2 (50 marks)

Paper 1 (10 marks)

1 The diagram below shows how a molecule might be biosynthesised from building blocks in a metabolic pathway.

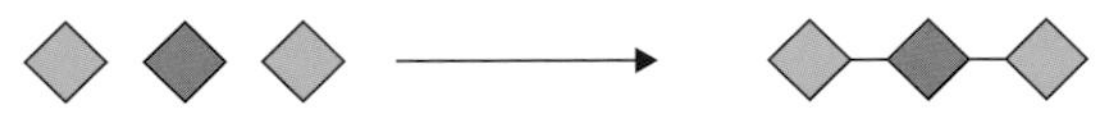

Which row in the table below correctly describes the metabolic process shown in the diagram and energy relationship involved in the reaction?

	Metabolic process	Energy relationship
A	Anabolic	Energy used
B	Anabolic	Energy released
C	Catabolic	Energy used
D	Catabolic	Energy released

2 The diagram on the right shows a metabolic pathway that is controlled by end product inhibition. For metabolite S to bring about end product inhibition, with which of the following must it interact?

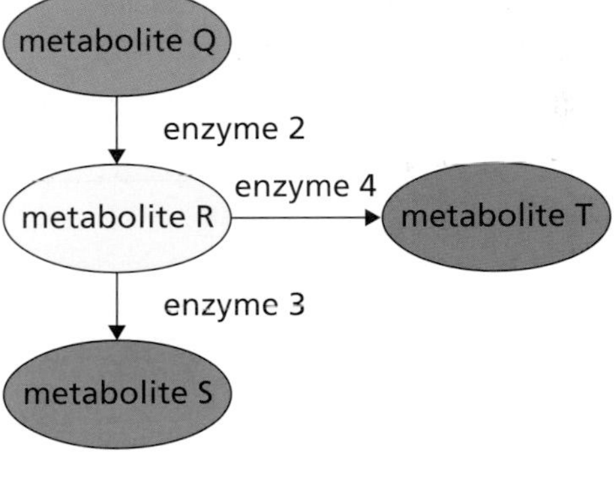

A metabolite P
B metabolite Q
C enzyme 1
D enzyme 4

3 In metabolic pathways, the rates of reaction can be affected by the presence of enzyme inhibitors.
Which row in the table below is correct?

	Type of inhibition	Inhibitor binds to active site	Effect of increasing substrate concentration on inhibition
A	competitive	yes	unaffected
B	non-competitive	no	unaffected
C	competitive	no	reversed
D	non-competitive	yes	reversed

4 During fermentation in muscle fibres, pyruvate is
A converted to citrate
B broken down by the mitochondria
C broken down to carbon dioxide and water
D converted to lactate.

5 One molecule of ATP stores 31 $kJmol^{-1}$ of energy.
During aerobic respiration, only 40 per cent of the energy content of glucose is used to generate ATP molecules.
How many ATP molecules would be generated from a glucose sample containing 8835 kJ energy?
A 114
B 285
C 3534
D 109 554

6 Which of the following abiotic factors does **not** affect an animal's ability to maintain its metabolic rate?
  A light intensity
  B temperature
  C salinity
  D pH

7 The South African giant bullfrog lives in a habitat in which drought conditions can occur at any time of the year. To survive, the bullfrog responds by becoming dormant. The name given to this type of dormancy is
  A predictive aestivation
  B predictive hibernation
  C consequential aestivation
  D consequential hibernation

8 A viable count of bacteria grown in batch culture is shown in the graph below.

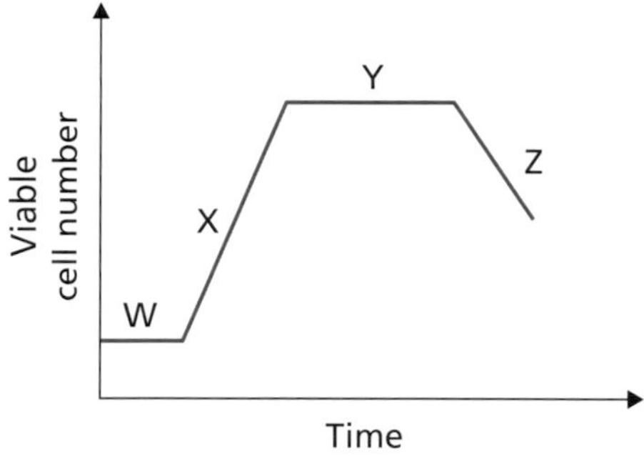

Four stages of growth are labelled W, X, Y and Z. Which row in the table below correctly identifies each stage of the bacterial growth curve?

	W	X	Y	Z
A	Log	Lag	Stationary	Death
B	Stationary	Log	Lag	Death
C	Lag	Log	Death	Stationary
D	Lag	Log	Stationary	Death

9 A bacterial culture contains 10 000 cells. If the doubling time for this bacterium is 30 minutes at 20 °C, how many cells will be present after 3 hours at 20 °C?
  A 60 000
  B 80 000
  C 320 000
  D 640 000

10 Yeast is often used as an alternative to bacteria as a recipient for foreign DNA because of yeast's ability to
  A grow rapidly in culture
  B produce active forms of the protein
  C undergo sexual reproduction
  D produce complex proteins.

Paper 2 (40 marks)

1 The following graph shows how the rate of an enzyme-catalysed reaction varies with the substrate concentration and how the reaction is affected by a competitive and a non-competitive inhibitor.

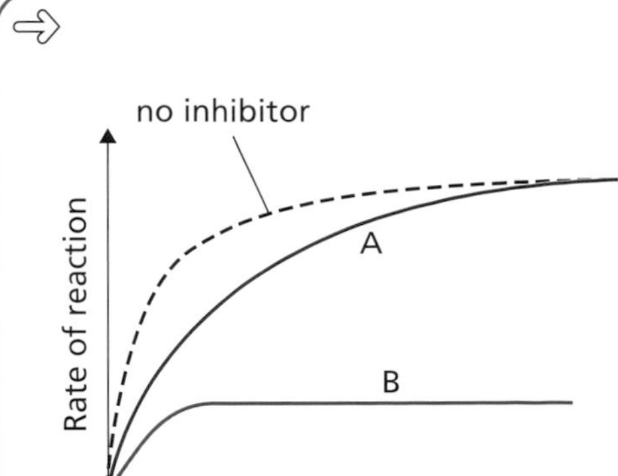

a) Explain the rate of reaction in the absence of an inhibitor. (2)
b) State which line shows the effect of the competitive inhibitor and explain the inhibitor's effect. (2)
c) Explain how a non-competitive inhibitor affects an enzyme. (1)

2 The diagram below shows part of the electron transport chain attached to the inner membrane of a mitochondrion.

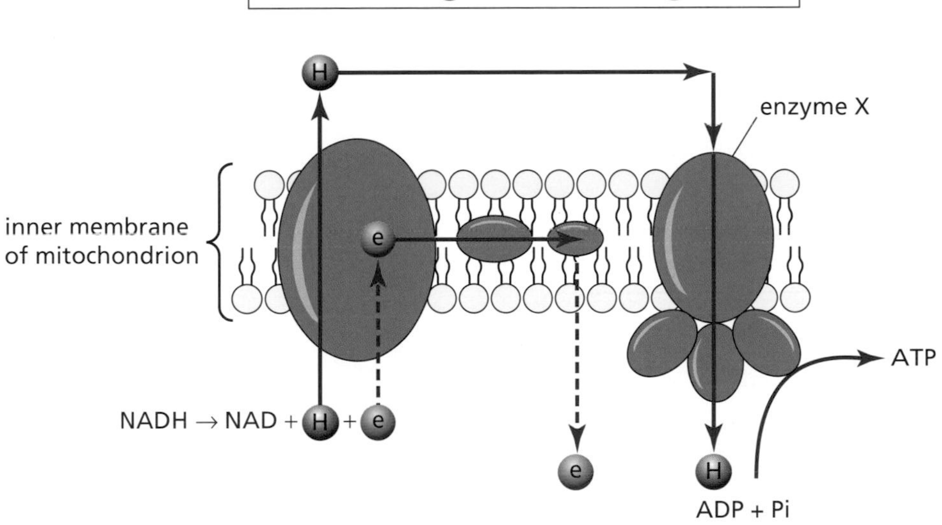

a) Describe the role of the electrons that pass down the chain. (1)
b) Describe the events involved in the synthesis of ATP by the electron transport chain. (2)
c) State the role of oxygen in the electron transport chain. (1)

Questions 3 and 4 each contain a choice.

3 Give an account of heart structure and circulatory system in
Either **A** amphibians
or **B** mammals (4)

4 Give an account of metabolism in relation to ecological niche in
Either **A** conformers
or **B** regulators (4)

5 The following diagram shows some stages in the control of body temperature in a mammal.

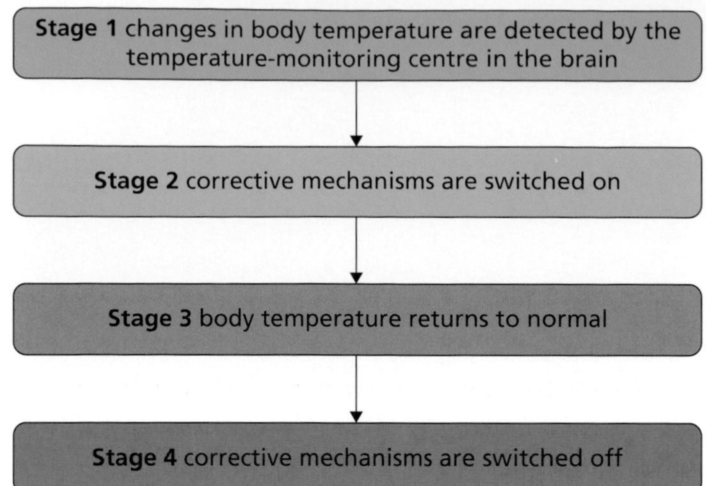

a) Name the temperature-monitoring centre in the brain. (1)

b) State the method by which instructions are passed from the temperature-monitoring centre to the effectors. (1)

c) Give **one** example of a corrective mechanism in response to an increase in temperature and explain how it would return the body temperature to normal. (2)

d) Explain how the mechanism of temperature control in the diagram demonstrates negative feedback. (2)

6 The graph below shows the growth of bacteria in a fermenter.

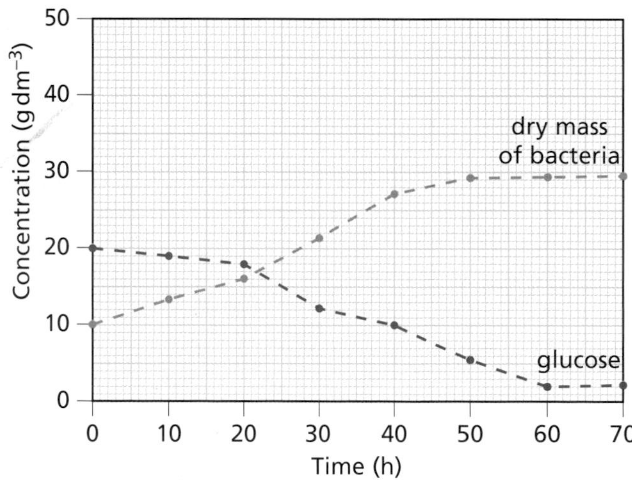

a) Explain why the fermenter is sterilised before adding the culture at the start of a procedure. (1)

b) State **two** other culture conditions that are controlled to provide optimum conditions for growth. (1)

c) Calculate the average increase per hour in the dry mass of the bacteria during the first 20-hour period. (1)

d) Explain the results obtained in the last 20 hours of the fermentation. (1)

7 One of the steps in the production of bacteria capable of producing human insulin involves inserting the gene that codes for human insulin into a plasmid vector.

The following diagram shows one of the genetically modified plasmid vectors.

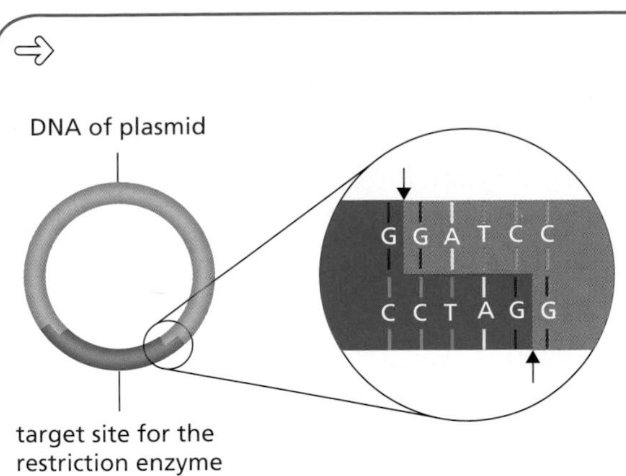

a) With reference to the diagram, explain the importance of using the same restriction enzyme to remove the insulin gene from a human chromosome and also to cut open the bacterial plasmid. (2)

b) Name the enzyme used to seal the insulin gene into the plasmid vector. (1)

c) Other than to make a desired protein, give **two** functions of gene sequences that are transferred to microorganisms by genetic engineering. (2)

Question 8 contains a choice.

8 *Either* **A** Give an account of aerobic respiration under the following headings:

a) glycolysis (2)

b) the citric acid cycle (6)

(total = 8)

or B Write notes on

a) adaptations shown by organisms to survive adverse conditions (5)

b) migratory behaviour (3)

(total = 8)

Answers to Practice course assessment: Area 2

Paper 1

1 A, 2 C, 3 B, 4 D, 5 A, 6 A, 7 C, 8 D, 9 D, 10 B [1 each = 10]

Paper 2

1 a) rate of reaction increases as substrate concentration increases because more substrate becomes available for available enzyme to work on/substrate concentration had been the limiting factor; rate of reaction then remains constant as all the enzyme molecules are being used/all available active sites are occupied/enzymes are now the limiting factor [1 each = 2]

b) A; as substrate concentration increases effect of inhibitor decreases [1 each = 2]

c) binds to the enzyme and changes the shape of the active site [1]

2 a) release/provide energy to pump H ions across the inner mitochondrial membrane [1]

b) return flow of H ions drives ATP synthase; ADP + Pi combine to form ATP [1 each = 2]

c) combines with hydrogen ions and electrons to form water [1]

3A Amphibians: incomplete double; heart with three chambers; made up of a right and left atrium and one lower ventricle; pressure maintained; but tissue blood is incompletely oxygenated [any 4 = 4]

B Mammals: complete double; heart has four chambers; made up of two atria and two ventricles; pressure maintained; and tissue blood completely oxygenated [any 4 = 4]

4 A Conformers: internal environment is dependent on their external environment; their metabolic costs may be low; conformers cannot alter their metabolic rate using physiological means; conformers may have a narrow ecological niche; behavioural responses may help to maintain optimum metabolic rate [any 4 = 4]

B Regulators: can control their internal environment and maintain a steady state; regulators adjust their metabolic rate using physiology; regulation requires energy to maintain a constant internal environment; increases the range of possible ecological niches that can be occupied; negative feedback is the control mechanism by which homeostasis is achieved [any 4 = 4]

5 a) hypothalamus [1]

b) nerve impulses [1]

c) vasodilation; increased blood flow to the skin increases heat loss [1 each = 2]
increased sweating; body heat is used to evaporate water in the sweat, cooling the skin [1 each = 2]

d) any change away from the norm is detected by receptor cells which switches on a corrective mechanism; once the conditions are returned to normal the corrective mechanism is switched off [1 each = 2]

6 a) to eliminate contaminating/competing microorganisms OR reduce competition with desired microorganisms for nutrients OR reduce the risk of spoilage of the product [any 1 = 1]

b) temperature; pH; oxygen [any 2 = 1]

c) 0.3 g dm^{-3} [1]

d) growth/dry mass of bacteria levels off/ stops as glucose/substrate decreases/runs out/is used up [1]

7 a) cuts DNA/chromosome *and* plasmid at certain base/target sequences; produces complementary base pairs or sticky ends that can then be combined [1 each = 2]

b) ligase [1]

c) sequences that prevent survival in external environment; sequences which are marker genes [1 each = 2]

8 A a) glycolysis is the breakdown of glucose to pyruvate; 2 ATP molecules are used to phosphorylate intermediates in glycolysis; energy investment phase; 4 ATP molecules are produced/generated/made in a pay-off stage [any 2 = 2]

b) if oxygen is available/in aerobic conditions pyruvate progresses to the citric acid cycle; pyruvate is converted/broken down to an acetyl group; acetyl group combines with coenzyme A; acetyl group combines with oxaloacetate to form citrate; citric acid cycle is enzyme-controlled/involves dehydrogenases; ATP generated/synthesised/produced/released at substrate level in the citric acid cycle; carbon dioxide is released from the citric acid cycle; oxaloacetate is regenerated; NAD/NADH transports electrons/ transports hydrogen ions to the electron transport chain [any 6 = 6]
[total = 8]

8 B a) some environments vary beyond tolerable limits; the extremes of conditions do not allow the normal metabolism of the organisms present; variation in conditions can be cyclical or unpredictable; metabolic rate can be reduced when conditions would make the cost of metabolic activity too high; dormancy may be predictive or consequential; example from hibernation/aestivation/leaf fall; daily torpor is a period of reduced activity; example of an organism and the adverse conditions that it survives
[any 5 = 5]

b) some organisms avoid adverse conditions by undertaking migration; migration uses energy to relocate to a more suitable environment; migratory behaviour can be innate and learned; example of techniques to track migration – satellite tracking OR leg rings (ringing and recapture) [any 3 = 3]
[total = 8]

Area 3 Sustainability and interdependence

Key Area 3.1a
Food supply

Key points

1. **Food security** is the ability of a human population to access food of sufficient quality and quantity. ☐
2. Increasing human population and concern for food security leads to a demand for increased food production. ☐
3. Food production must be sustainable and not degrade the natural resources on which **agriculture** depends. ☐
4. Agricultural production depends on factors that control **photosynthesis** and plant growth. ☐
5. The area to grow crops is a limiting factor. ☐
6. Increased food production will depend on factors that control plant growth including the breeding of higher yielding **cultivars**, the use of **fertiliser**, and protecting crops from **pests**, diseases and **competition**. ☐
7. All food production is dependent ultimately on photosynthesis. ☐
8. Plant crop examples include cereals, potato, **root crops** and **legumes**. ☐
9. Breeders seek to develop crops with higher nutritional values, resistance to pests and diseases, physical characteristics suited to rearing and harvesting as well as those that can thrive in particular environmental conditions. ☐
10. **Livestock** produce less food per unit area than crop plants, due to loss of energy between **trophic levels**. ☐
11. Livestock production is often possible in habitats that are unsuitable for growing crops. ☐

Summary notes

Food security and agricultural production

Food security is the ability of a human population to access food of sufficient quality and quantity over a sustained period to avoid starvation or malnutrition.

The following table shows some aspects that underpin food security.

Aspect of food security	Notes
Availability	Existence of food in sufficient quantities and of appropriate quality (nutritional value)
Accessibility	Sufficient economic and infrastructure resources to access available food resources
Usage	Level of nutritional knowledge needed to use food resources properly
Sustainability	Degree to which food security can be guaranteed over extended time periods

An increase in the global human population has raised the overall requirement for food and increased the difficulty in ensuring food security for all. Figure 3.1 shows the increase in the human population over the last 2000 years. It is estimated that possibly 30 per cent of the current population of 7 billion humans lack sufficient food security and so are liable to starvation or malnutrition.

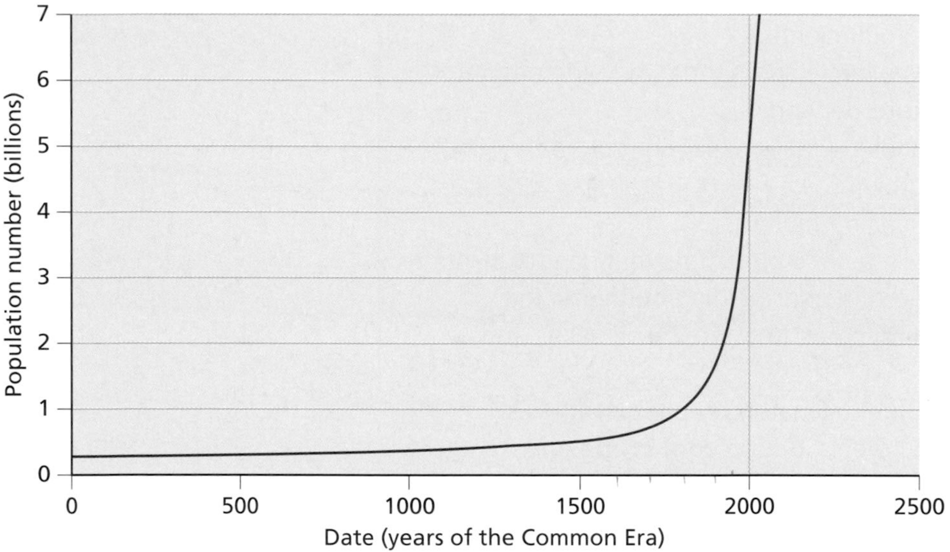

Figure 3.1 Human population growth

Energy from the Sun drives all of the Earth's food chains. Green plants trap light energy in photosynthesis. Some of this is passed to animal consumers. Humans can occupy different trophic levels, as shown in the energy pyramids in Figure 3.2. Due to the loss of energy between trophic levels, it is more energy-efficient for humans to act as primary consumers and eat crops rather than eating livestock as secondary consumers.

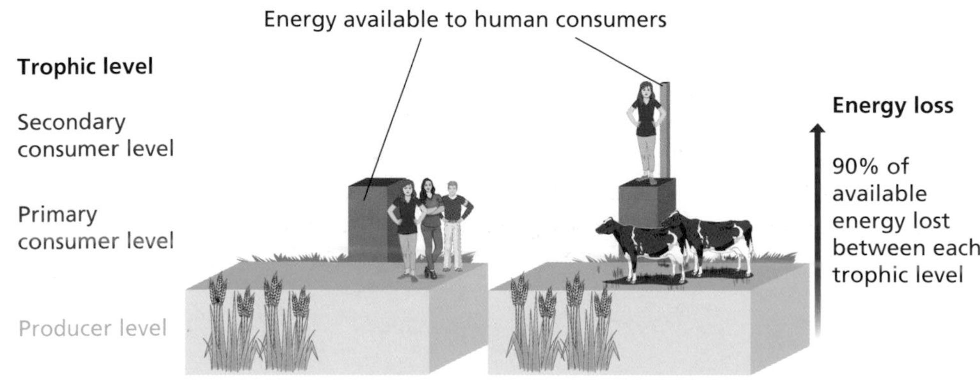

Figure 3.2 Humans and the pyramid of energy

Agriculture

Food production must be sustainable and not degrade natural resources, such as the soil, on which agriculture depends. Agricultural production depends on factors that control photosynthesis and plant growth. The area to grow crops is a limiting factor. Increased food production depends on factors that control plant growth, including the breeding of higher yielding cultivars, the use of fertiliser, and protecting crops from pests, diseases and competition.

All food production ultimately depends on photosynthesis.

Crop plants

A small number of green plant species, such as sugar cane, cereal crops, potatoes and legumes, are the main producers of human food, as shown in the following table.

Crop	Average world production (millions of tonnes per annum)
Sugar cane	1800
Maize (cereal crop)	885
Rice (cereal crop)	723
Wheat (cereal crop)	701
Potato	373
Soya bean (legume)	262

Crop production depends on a large number of factors, as shown in the following table. The environmental factors are mainly those that affect photosynthesis. Production can be increased by controlling these factors.

Soil factors	Environmental factors	Crop factors
Area of soil under crop	Temperature	Cultivar of crop
Fertiliser levels in soil	Rainfall	Pests
Soil type	Atmospheric carbon dioxide levels	Disease
Drainage	Light availability	Competition

Breeders seek to develop crops with higher nutritional values, resistance to pests and diseases, physical characteristics suited to rearing and harvesting, as well as those that can thrive in particular environmental conditions.

Livestock animals

Since energy is lost between trophic levels, keeping animal livestock can result in less food production per unit area of land in agriculture. Land planted with a cereal crop such as wheat can be more productive in terms of overall food than the same area of pasture land kept for cattle. Livestock production is often possible in habitats unsuitable for growing crops, such as hill farming in upland regions.

Key words

Agriculture – human practice of growing crops and keeping livestock to maintain food security
Competition – struggle for existence between two organisms
Cultivar – varieties of cultivated crops, for example high yielding, disease resistant, genetically modified (GM) cultivars
Fertiliser – chemical addition to soil to increase plant growth
Food security – the ability to access sufficient quantity and quality of food over a sustained period of time
Legume – plant with seeds in a pod, such as the bean or pea
Livestock – agricultural animals
Pest – organisms such as insects, nematode worms and molluscs, which damage agriculture and reduce food security
Photosynthesis – production of carbohydrate by a plant using the energy of light
Root crop – a crop that is a root vegetable, for example carrot or sugar beet
Trophic level – feeding level in a food chain

Questions ?

Short answer (1 or 2 marks)

1 Explain why there has been a major increase in concern about food security in recent years. (1)
2 State what is meant by the following terms:
   a) photosynthesis
   b) trophic level (2)
3 Give **three** examples of methods of increasing yield of food from a crop plant such as wheat. (2)
4 Explain why the yield of food from cultivated crops would be greater than the yield of meat from livestock kept in an equivalent area of land. (2)

Longer answer (3–10 marks)

5 Give an account of food security under the following headings:
   a) availability and accessibility (3)
   b) usage and sustainability (2)
(total = 5)

Answers are on page 136.

Key Area 3.1b

Photosynthesis

Key points 

1 Photosynthesis captures **light energy** to produce carbohydrates. ☐
2 Light energy is absorbed by photosynthetic **pigments** to generate ATP and for **photolysis**. ☐
3 Light energy that is not absorbed is **transmitted** or **reflected**. ☐
4 The **absorption spectrum** shows the range and extent to which light wavelengths are absorbed by a particular pigment. ☐
5 The **action spectrum** shows the rate of photosynthesis of a plant across a range of light wavelengths. ☐
6 **Chlorophyll** is the main photosynthetic pigment in green plants. ☐
7 **Carotenoids** extend the range of wavelengths absorbed and pass the energy to chlorophyll for photosynthesis. ☐
8 Each pigment absorbs a different range of wavelengths of light. ☐
9 Absorbed light energy excites electrons in the pigment molecule. ☐
10 Transfer of electrons through the electron transport chain in the membrane of the chloroplast releases energy to generate ATP by ATP synthase. ☐
11 Energy is also used for photolysis, in which water is split into oxygen, which is evolved, and hydrogen, which is transferred by the **coenzyme NADP** to the **carbon fixation stage** (Calvin cycle). ☐
12 In the carbon fixation stage (Calvin cycle), the enzyme **RuBisCO** fixes carbon dioxide by attaching it to **ribulose bisphosphate (RuBP)** to produce an intermediate called **3-phosphoglycerate (3PG)**. ☐
13 The 3-phosphoglycerate (3PG) is phosphorylated by ATP and combined with hydrogen from NADPH to form **glyceraldehyde-3-phosphate (G3P)**. ☐
14 G3P is used to regenerate RuBP and for the synthesis of glucose.
15 Glucose may be used as a respiratory substrate, synthesised into **starch** or **cellulose** or passed to other biosynthetic pathways. ☐
16 These biosynthetic pathways can lead to the formation of a variety of metabolites such as DNA, protein and fat. ☐

Summary notes

Photosynthesis

Photosynthesis is a process in which green plants trap light energy and use it in the production of carbohydrate. Carbon dioxide from the atmosphere is combined with water from the soil to make glucose. Oxygen is produced and released as a by-product. Photosynthesis is summarised in Figure 3.3.

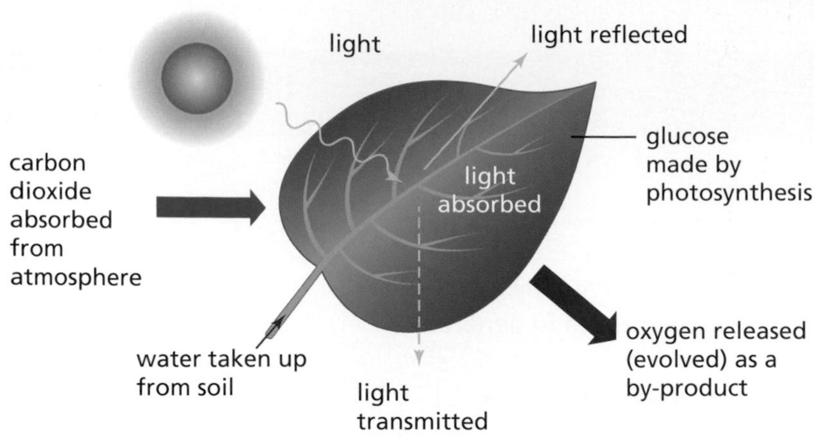

Figure 3.3 Summary of photosynthesis

Hints & tips

Remember ***ART*** *–* ***a****bsorbed,* ***r****eflected and* ***t****ransmitted.*

Light energy and pigments

Light energy travels in waves. Green plants have several pigments that are able to absorb light energy. Chlorophyll a and b and carotenoids are the commonly occurring pigments. Paper or thin layer chromatography can be used to separate the photosynthetic pigments in the chloroplast of a leaf. This is a technique which you need to know for your exam. Light that strikes pigment molecules is absorbed, transmitted or reflected, as shown in Figure 3.3 above.

T

The absorption spectrum is a graph that shows the range and extent to which each wavelength of light is absorbed by a particular pigment. Figure 3.4 shows the absorption spectrum for chlorophyll a. The absorption spectrum can be observed by passing light through a pigment extract, then through a spectroscope. Use of a spectroscope is a technique you need to be familiar with for your exam. The action spectrum shows the rate of photosynthesis at each wavelength of light in a plant, as shown in Figure 3.5.

T

Key links

There is an example of a question on using chromatography in the Skills of scientific inquiry chapter on page 146.

Key links

There is an example of a question on using a spectroscope in the Skills of scientific inquiry chapter on page 146.

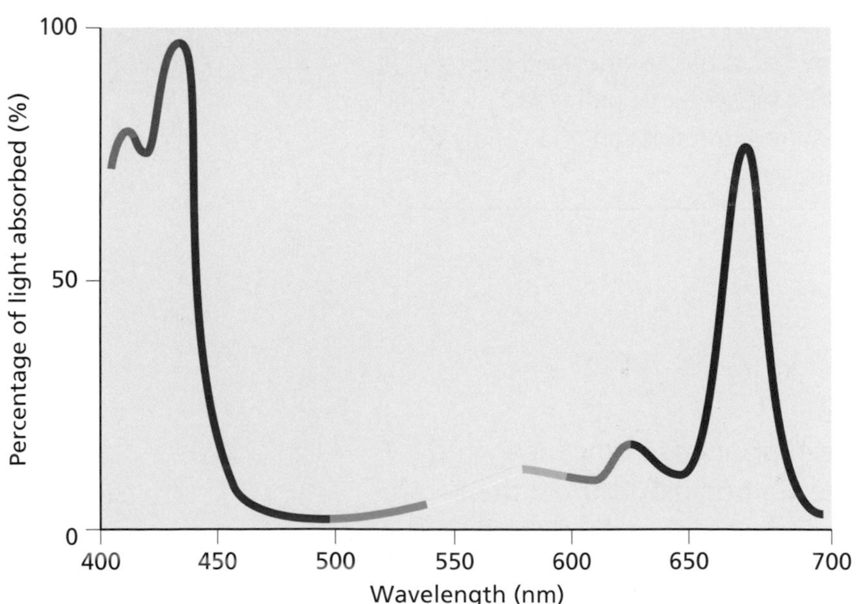

Figure 3.4 Absorption spectrum of chlorophyll a

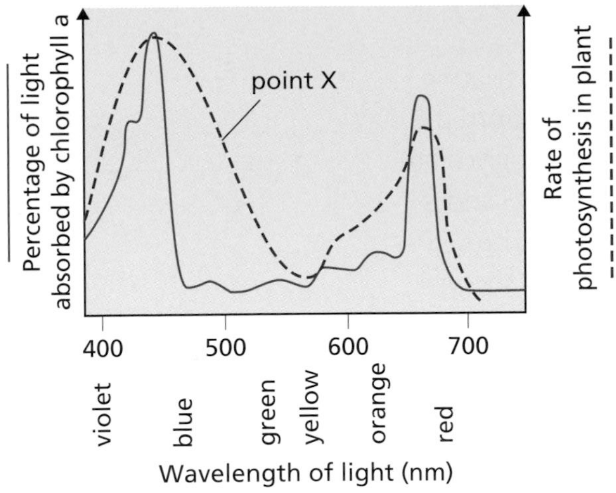

Figure 3.5 Absorption spectrum for chlorophyll a compared with the action spectrum for a green plant

Notice how the absorption spectrum and the action spectrum are broadly similar, though not identical. Comparison of the two spectra helps to confirm that pigments other than chlorophyll a are used in photosynthesis. At point X on the graph, the rate of photosynthesis remains high although the absorption by chlorophyll a is low.

Chlorophyll a is the main photosynthetic pigment in green plants. Carotenoids extend the range of wavelengths absorbed and pass the energy to chlorophyll for photosynthesis. Each pigment absorbs a different range of wavelengths of light.

Energy capture

When pigments absorb light, the energy excites electrons in the pigment molecules. These electrons move through a series of electron carrier molecules attached to the membranes of chloroplasts, releasing their energy, which is then used by ATP synthase to generate ATP from ADP + Pi. Some energy is also used in photolysis to split water into oxygen, which is released (evolved), and hydrogen, which becomes bound to the coenzyme NADP, as shown in Figure 3.6. The ATP and NADPH are passed to the next stage, the carbon fixation stage (Calvin cycle).

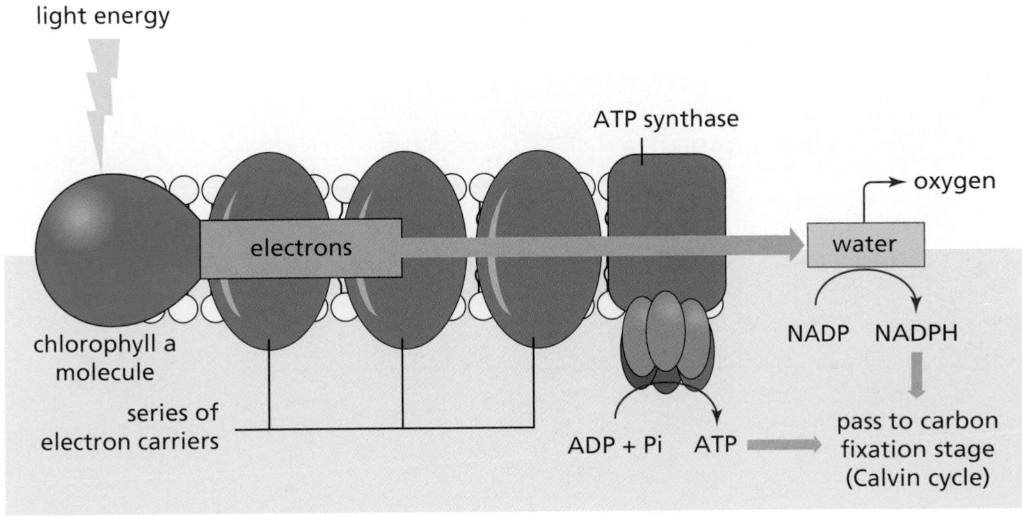

Figure 3.6 Light-dependent stage of photosynthesis on a chloroplast membrane

Carbon fixation stage (Calvin cycle)

Carbohydrate is produced by a metabolic pathway called the carbon fixation stage (Calvin cycle). The enzyme RuBisCo fixes carbon dioxide (CO_2) from the air by attaching it to ribulose bisphosphate (RuBP) to form 3-phosphoglycerate (3PG). The 3-phosphoglycerate accepts (3PG) hydrogen from NADPH and is phosphorylated by ATP to form glyceraldehyde-3-phosphate (G3P). The G3P is then synthesised into glucose or can be used to regenerate RuBP, as shown in Figure 3.7.

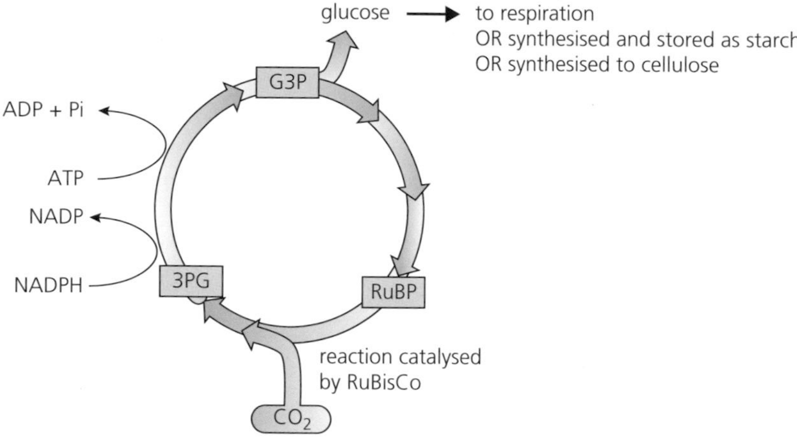

Figure 3.7 Carbon fixation stage (Calvin cycle)

The glucose produced can be used as a respiratory substrate to drive the plant's life processes or can be synthesised to cellulose or starch. Other carbohydrate formed in photosynthesis can be used in the synthesis of a variety of different metabolites, including DNA, protein and fat.

Figure 3.8 shows a summary of the stages of photosynthesis.

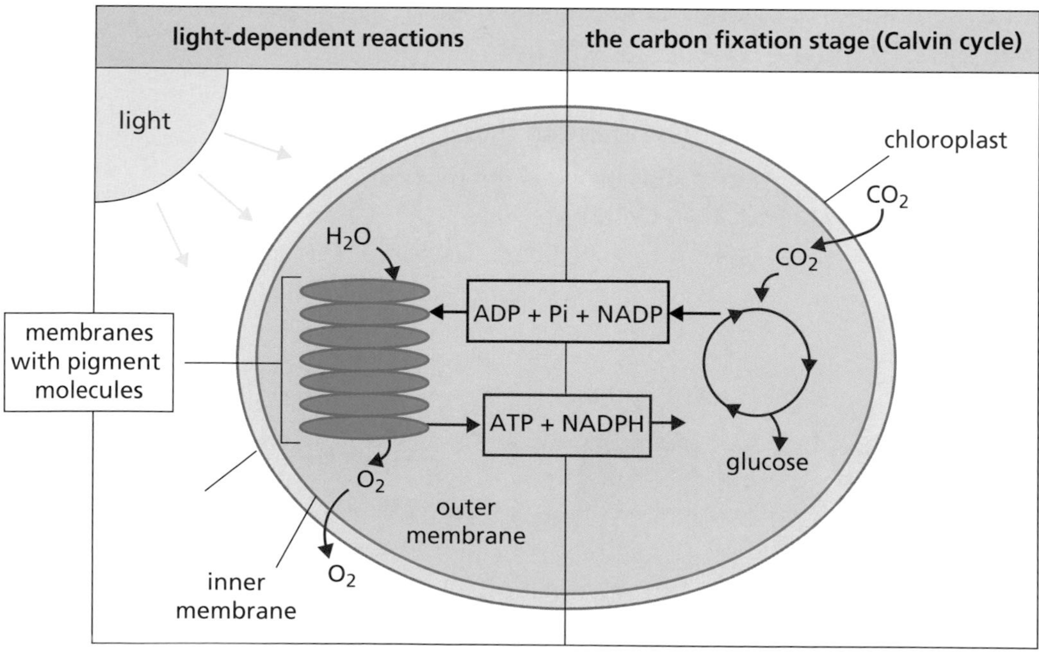

Figure 3.8 The two stages of photosynthesis summarised

Key words

3-phosphoglycerate (3PG) – produced when CO_2 is fixed to RuBP by RuBisCo
Absorption spectrum – graph showing wavelengths of light absorbed by a pigment
Action spectrum – the rate of photosynthesis carried out at each wavelength of light
Carbon fixation stage – Calvin cycle; second stage in photosynthesis, which results in the production of glucose
Carotenoids – orange and yellow accessory pigments in plants that extend the range of wavelengths absorbed and pass the energy to chlorophyll for photosynthesis
Cellulose – structural carbohydrate in cell walls derived from photosynthesis
Chlorophyll – green pigment molecule in plants that absorbs red and blue light for photosynthesis
Coenzyme NADP – hydrogen carrier in photosynthesis
Glyceraldehyde-3-phosphate (G3P) – compound in the carbon fixation stage that can be converted to glucose or used to regenerate RuBP
Light energy – radiant energy used in photosynthesis
Photolysis – light energy splits water into oxygen, which is evolved, and hydrogen, which is transferred to the coenzyme NADP
Pigment – coloured substance that absorbs light for photosynthesis
Reflection – light that strikes a leaf passes away from its surface back to the atmosphere
Ribulose bisphosphate (RuBP) – the carbon dioxide acceptor molecule in the carbon fixation stage
RuBisCo – the enzyme that fixes carbon dioxide by attaching it to RuBP
Starch – storage carbohydrate in plants
Transmission – physical process of passing light energy through a surface

Questions ?

Short answer (1 or 2 marks)

1 State what can happen to light that strikes a green leaf. (2)
2 Explain the difference between the absorption spectrum of a pigment and the action spectrum of a green plant. (2)
3 Describe the part played by carotenoid pigments in photosynthesis. (2)
4 Name the process which involves the breakdown of water during the light-dependent stage of photosynthesis. (1)
5 Describe the role of the electrons that are released when light strikes a chlorophyll molecule. (2)
6 Name the enzyme that joins carbon dioxide to ribulose bisphosphate in the carbon fixation stage. (1)
7 Name the coenzyme that transports hydrogen to the carbon fixation stage. (1)
8 Describe the fate of G3P produced during the carbon fixation stage. (2)

Longer answer (3–10 marks)

T 9 Describe how an absorption spectrum for a plant pigment extract can be observed. (2)
10 Describe the role of the photosynthetic pigments in the chloroplast. (6)
T 11 Describe a technique that is used to separate the different photosynthetic pigments in a green plant. (3)
12 Give an account of photosynthesis under the following headings:
a) the capture of light energy (5)
b) the carbon fixation stage (Calvin cycle) (4)
(total = 9)

Answers are on page 136.

Key Area 3.2

Plant and animal breeding

Key points

1 Plant and animal breeding is carried out to improve characteristics to help support sustainable food production. ☐

2 Breeders develop crops and animals with higher food yields, higher nutritional values, pest and disease resistance and ability to thrive in particular environmental conditions. ☐

3 Plant **field trials** are carried out in a range of environments to compare the performance of different cultivars or treatments and to evaluate the performance of **genetically modified (GM) crops**. ☐

4 In designing field trials, account has to be taken of the **selection of treatments**, the number of **replicates** and the **randomisation** of treatments. ☐

4 The selection of treatments must ensure valid comparisons between cultivars, the number of replicates involved must take account of the variability within the sample, and the randomisation of treatments is needed to eliminate bias when measuring treatment effects. ☐

6 In **inbreeding**, selected related plants or animals are bred for several generations until the population breeds true to the desired type due to the elimination of **heterozygotes**. ☐

7 Inbreeding can result in an increase in the frequency of individuals who are **homozygous** for **recessive** deleterious alleles. ☐

8 These individuals are less successful at surviving to reproduce, resulting in **inbreeding depression**. ☐

9 In animals, individuals from different breeds may produce a new **cross-breed population** with improved characteristics. ☐

10 The two parent breeds can be maintained to produce more cross-bred animals showing the improved characteristic. ☐

11 New alleles can be introduced to plant and animal lines by crossing a cultivar or breed with an individual with a different, desired genotype. ☐

12 In plants, **F_1** hybrids, produced by the crossing of two different inbred lines, create a relatively uniform heterozygous crop. ☐

13 F_1 hybrids often have increased vigour and yield. ☐

14 Plants with increased **hybrid vigour** may have increased disease resistance or increased growth rate. ☐

15 In inbreeding animals and plants, F_1 hybrids are not usually bred together as the **F_2** generation shows too much variation. ☐

16 As a result of genome sequencing, organisms with desirable genes can be identified and then used in breeding programmes. ☐

17 Single genes for desirable characteristics can be inserted into the genomes of crop plants, creating genetically modified plants with improved characteristics. ☐

⇨

18 Breeding programmes can involve crop plants that have been genetically modified using recombinant DNA technology. ☐

19 **Recombinant DNA technology** in plant breeding includes insertion of the **Bt toxin gene** into plants for pest resistance and the insertion of the **glyphosate resistance gene** for herbicide tolerance. ☐

Summary notes

Breeding programmes

One method of increasing food security is to develop new varieties of crops and livestock breeds using breeding programmes. These programmes allow desirable features to be bred into particular plants or animals. Desirable features include higher yields, higher nutritional values, pest and disease resistance, the ability to thrive in particular environmental conditions, and characteristics that assist rearing and harvesting. The following table shows some of the desirable characteristics that have been bred into plant crops and animal breeds.

Plant crop or animal breed	Desirable characteristic that increases food security
Wheat	High grain yield
Potato	Resistance to fungal disease
Soya bean	High protein content of seeds
Strawberry	Resistance to frost
Dairy cattle	High milk yield
Beef cattle	High meat yield

Dwarf varieties

Fifty years ago, plant breeders discovered dwarf varieties of some cereal crops with much shorter stems than normal. These plants put more of their energy into creating seed and were easier to harvest, thus increasing the yield of these crops and improving food security.

Selecting and breeding

Plant and animal breeders work to cross individuals with desired characteristics such as high milk yielding cattle or dwarf crop plants.

Inbreeding and its results

In inbreeding, selected related plants or animals are bred for several generations until the population breeds true to the desired type due to the elimination of heterozygotes. Inbreeding ensures that offspring receive the alleles desired and eventually form a homozygous stock that will continue to breed true for the desired characteristic over many generations.

A disadvantage of inbreeding is that it can result in an increase in the frequency of individuals who are homozygous for recessive deleterious alleles. These individuals are less successful at surviving to reproduce. This effect is known as inbreeding depression. The following table shows how repeated inbreeding increases the percentage of homozygous offspring.

Generation	P	F_1 ratio	F_2 ratio	F_3 ratio
Genotypes	**Aa** (self-pollinated)	1**AA**:2**Aa**:1**aa** (all allowed to self-pollinate)	6**AA**:4**Aa**:6**aa** (all allowed to self-pollinate freely)	28**AA**:8**Aa**:28**aa**
Percentage of generation that are homozygous	0%	2 out of 4 = 50%	12 out of 16 = 75%	56 out of 64 = 87.5%

Cross-breeding and F_1 hybrids

In animals, individuals from different breeds may produce a new cross-breed population with improved characteristics. The two parent breeds can be maintained to produce more cross-bred animals showing the improved characteristic. New alleles can be introduced to plant and animal lines by crossing a cultivar or breed with an individual with a different, desired genotype.

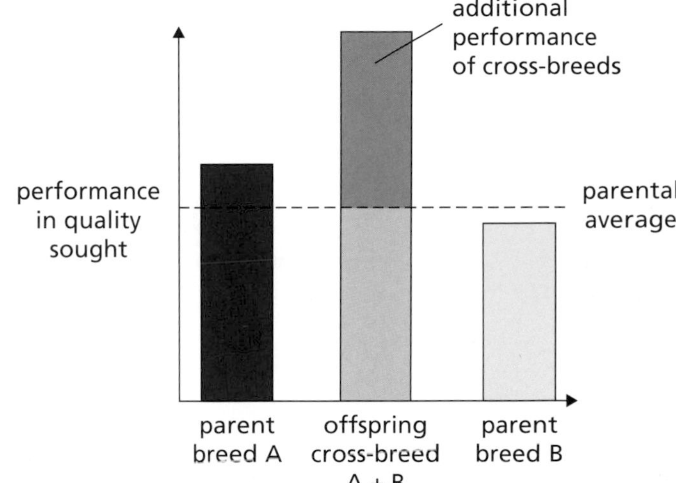

Figure 3.9 Results of cross-breeding two pig varieties

Breeders often deliberately maintain stock by cross-breeding regularly. The F_1 individuals produced often show increased vigour and yield because they combine the qualities of their parents. In plants, F_1 hybrids, produced by the crossing of two different inbred lines, create a relatively uniform heterozygous crop. Plants with increased vigour may have increased disease resistance or increased growth rate. Figure 3.9 shows how hybrid vigour is obtained by cross-breeding pig varieties A and B.

The F_1 performs well in the quality sought, but in most cases it is used for food because further breeding of hybrids results in varied offspring, some of which would be totally unsuitable. In inbreeding animals and plants, F_1 hybrids are not usually bred together as the F_2 generation shows too much variation.

Genetic technology

Organisms can have their genome sequenced and those with desirable alleles can be identified and used in breeding programmes to improve stock.

Genetic technology allows genetic material from one species to be inserted into the genome of another. The transformed organisms can then be used in breeding programmes. Single genes for desirable characteristics can be inserted into the genomes of crop plants, creating genetically modified plants with improved characteristics. The potential benefits are huge because the genetic material of the transformed organism is undisturbed apart from the insertion of a gene.

Key links

There is more about gene sequencing in Key Area 1.8 (page 32).

In many cases the transformation involves adding genetic material to a crop plant genome by infecting the plant with bacteria which have plasmids modified to contain a desired gene. The desired gene is passed horizontally into the crop plant genome directly from the bacterial cell plasmids (see Figure 1.27(b) on page 27). Recombinant DNA technology in plant breeding includes insertion of the Bt toxin gene into plants for pest resistance and the insertion of the glyphosate resistance gene for herbicide tolerance.

Key links

There is more about horizontal gene transfer in Key Area 1.7 (page 26).

The following table shows examples of how genetic transformation techniques have been used very successfully in recent years.

Crop plant transformed	Genetic material added	Food security benefit
Rice	Genes for vitamin A	Rice plants providing better nutrition for human consumers
Maize	Gene for Bt toxin, which kills insects	Maize plants resistant to insect pests, so increasing yield
Soya bean	Gene for glyphosate resistance	Soya fields can be sprayed with herbicide, killing weeds without damaging the crop and so increasing crop yield

Field trials

Plant field trials are carried out in a range of environments to compare the performance of different cultivars or treatments and to evaluate the performance of GM crops. In designing field trials, account has to be taken of the selection of treatments, the number of replicates and the randomisation of treatments. The selection of treatments must ensure valid comparisons between cultivars, the number of replicates involved must take account of the variability within the sample, and the randomisation of treatments is needed to eliminate bias when measuring treatment effects. Figure 3.10 shows how field trials on different fertiliser treatments and different cultivars could be set up.

TRIAL X

1 kg of cultivar A seed and 1 kg fertiliser per hectare	1 kg of cultivar A seed and 2 kg fertiliser per hectare
1 kg of cultivar A seed and 3 kg fertiliser per hectare	1 kg of cultivar A seed and 4 kg fertiliser per hectare

TRIAL Y

1 kg of cultivar A seed and 1 kg fertiliser per hectare	1 kg of cultivar B seed and 1 kg fertiliser per hectare
1 kg of cultivar C seed and 1 kg fertiliser per hectare	1 kg of cultivar D seed and 1 kg fertiliser per hectare

Figure 3.10 Diagram of field trials X and Y

Trial X shows four field plots with the *same* cultivar at *different* fertiliser treatments but all other factors held constant.

Trial Y shows four field plots with *different* cultivars at the *same* fertiliser treatment but all other factors held constant.

The plots can be harvested after a set time and yields compared. There are some flaws in these procedures. In order to ensure that the trials are fair, replication and randomisation of treatments should be carried out. Figure 3.11 shows an improved version of trial X.

1 kg of cultivar A seed and 2 kg fertiliser per hectare	1 kg of cultivar A seed and 4 kg fertiliser per hectare	1 kg of cultivar A seed and 4 kg fertiliser per hectare	1 kg of cultivar A seed and 3 kg fertiliser per hectare
1 kg of cultivar A seed and 1 kg fertiliser per hectare	1 kg of cultivar A seed and 2 kg fertiliser per hectare	1 kg of cultivar A seed and 3 kg fertiliser per hectare	1 kg of cultivar A seed and 1 kg fertiliser per hectare

Figure 3.11 Diagram of an improved field trial X with plots replicated and randomised

Notice that each plot is of the same size and planted with the same mass and cultivar of seed, so that a fair comparison is ensured. The fertiliser treatments have been replicated so that variation in the individual seed samples can be eliminated. The positioning of the plots is randomised to eliminate bias created by the environment when measuring the effects of the treatments. This version could be carried out again with each of the four cultivars in trial Y.

Key words

Bt toxin gene – a gene that is inserted into plants using recombinant DNA technology to produce a protein that acts as a pesticide
Cross-breed population – population showing improved characteristics, produced by crossing individuals from different breeds
F_1 – first generation of offspring from a genetic cross
F_2 – offspring of an F_1 generation
Field trials – non-laboratory tests carried out in a range of environments to compare the performance of different cultivars or treatments and to evaluate GM crops
Glyphosate resistance gene – a gene that is inserted into plants using recombinant DNA technology to provide herbicide tolerance
GM crop – genetically modified crop which contains a gene from another species
Heterozygous – having two different alleles of the same gene and so not true breeding
Homozygous – having two identical alleles of the same gene and so true breeding
Hybrid vigour – the increase in such characteristics as size, growth rate, fertility, and yield of a hybrid organism over those of its parents
Inbreeding – crossing organisms of the same or similar genotype for several generations until the population breeds true for the desired characteristics
Inbreeding depression – an increase in the frequency of individuals who are homozygous for recessive deleterious alleles which lower biological fitness
Randomisation – a methodology based on chance, used to eliminate bias when measuring treatment effects
Recessive – alleles which only show in the phenotype when they are in homozygous form
Recombinant DNA technology – single genes for desirable characteristics can be inserted into the genomes of crop plants, creating genetically modified plants with improved characteristics
Replicates – field trials should be repeated to take account of variability within the sample
Selection of treatments – choice of treatments in a field trial such that a valid comparison can be made

Questions ?

Short answer (1 or 2 marks)

1 Give two desirable qualities that might be selected by breeders who are seeking to improve a crop plant species. (2)

2 Describe the role of inbreeding in the breeding of livestock. (2)

3 Explain the meaning of the term *inbreeding depression*. (2)

4 Explain why individuals from different breeds are sometimes crossed. (1)

5 In plant breeding, F_1 hybrids are often produced because they combine desired features of their parent varieties.
Explain why F_2 plants produced from the hybrid are considered of little use for further production. (1)

6 Name the procedure carried out to identify organisms with desirable genes for use in breeding programmes. (1)

Longer answer (3–10 marks)

7 Give an account of the benefits of using recombinant DNA technology in crop plants. (6)

Answers are on page 137.

Key Areas 3.3 and 3.4

Crop protection and animal welfare

Key points

1. Weeds compete with crop plants, while other pests and diseases damage them, and all reduce productivity. ☐
2. Properties of **annual weeds** include rapid growth, short life cycle, high seed output and long-term seed viability. ☐
3. Properties of **perennial weeds** with competitive adaptations include storage organs and **vegetative reproduction**. ☐
4. Most **crop plant pests** are invertebrate animals such as insects, nematode worms and molluscs. ☐
5. Plant diseases can be caused by fungi, bacteria or viruses, which are often carried by invertebrates. ☐
6. Weeds, other pests and diseases can be controlled by **cultural** methods that include ploughing, weeding and crop rotation. ☐
7. **Pesticides** include herbicides to kill weeds, **fungicides** to control fungal diseases, insecticides to kill insect pests, molluscicides to kill mollusc pests and nematicides to kill nematode pests. ☐
8. **Selective herbicides** have a greater effect on certain plant species (broad-leaved weeds). ☐
9. **Systemic herbicide** spreads through the vascular systems of plants and prevents regrowth. ☐
10. Systemic insecticides, molluscicides and nematicides spread through the vascular systems of plants and kill pests which are feeding on the plants. ☐
11. Problems with pesticides include toxicity to non-target species, **persistence** in the environment, **bioaccumulation** or **biomagnification** in food chains and the production of resistant populations of pests. ☐
12. Bioaccumulation is a build-up of a chemical in an organism. ☐
13. Biomagnification is an increase in the concentration of a chemical between trophic levels. ☐
14. Applications of fungicide based on **disease forecasts** are more effective than treating diseased crops. ☐
15. Weeds, other pests and diseases can also be controlled by using biological control and **integrated pest management (IPM)**. ☐
16. In biological control, the control agent is a natural predator, parasite or pathogen of the pest. ☐
17. IPM is a combination of chemical, biological and cultural control. ☐
18. A risk associated with **biological control** is that the control organism may become an invasive species, parasitise, prey on or be a pathogen of other species. ☐
19. There are costs, benefits and ethics associated with providing different levels of **animal welfare** in livestock production. ☐
20. Intensive farming is less ethical than free-range farming due to poorer animal welfare. ☐

⇨

21 Free-range farming requires more land and is more labour intensive but free-range goods can be sold at a higher price and animals raised in this way have a better quality of life. ☐

22 Intensive farming often creates conditions of poor animal welfare but is usually more cost effective, generating higher profit, as costs are lower. ☐

23 Behavioural indicators of poor animal welfare are **stereotypy**, **misdirected behaviour** and failure in sexual or parental behaviour. ☐

24 Very low (apathy) or very high (hysteria) levels of activity are also behavioural indicators of poor animal welfare. ☐

Summary notes

Agricultural ecosystems

Agricultural ecosystems are often very uniform, with only one species, the crop, making up the bulk of the community. This type of situation provides opportunities for weed plants, pests and diseases, which are liable to affect the crop plants and reduce yield. Weeds compete with crop plants, while other pests and diseases damage them, and all reduce productivity. A variety of methods of control are used to protect crops and reduce impact on food security.

Crop plant weeds

Crop weeds grow among crop plants where they reduce yields and productivity by competing for resources such as light, water, minerals and root space. Weeds can be annual plants or perennial plants. The following table defines and lists the main properties of annual and perennial weeds.

Type of weed	Definition	Properties
Annual	Grow from seed and complete their life cycle in one year	Grow quickly following germination Have a short life cycle High seed output Long-term seed viability
Perennial	Persist from year to year	Have competitive advantage through being established prior to crop growth Have storage organs Reproduce vegetatively using special structures such as runners and bulbs

Invertebrate pests

Most of the pests of crop plants come from three main invertebrate groups. The following table shows the main groups with notes on their adaptations.

Invertebrate group	Example(s)	Adaptations
Nematodes (tiny worms)	Eel worms	Bore into host plant and live parasitically within the plant tissues
Molluscs	Slugs, snails	Have rasping mouthparts, which can deal with tough plant material
Insects	Greenfly, caterpillars	Have piercing or biting mouthparts, which penetrate or chew plant tissues

Key links 

There is more about parasites in Key Area 3.5 (page 124).

Diseases of crop plants

Many crop plant diseases are caused by microorganisms such as fungi, bacteria and viruses, as shown in the following table. Some of these microorganisms can be carried by invertebrates acting as vectors.

Key links 

There is more about vectors in Key Area 3.5 (page 124).

Microorganism	Example	Affected crop plant
Fungus	Blight fungus	Potato
Bacterium	Soft rot	Parsley
Virus	Stunt virus	Tomato

Fungal diseases of crop plants often spread quickly and do huge damage rapidly. Protective applications of fungicide based on disease forecasts are often more effective than trying to treat an already diseased crop.

Control methods

Cultural methods

Cultural methods used in the control of weeds, other pests and diseases are often traditional and preventative and have been embedded in farming practices over many generations. The following table shows some examples and their effects.

Cultural practice	Effect
Ploughing	Perennial weeds damaged or buried
Weeding	Early hoeing removes annual weeds
Crop rotation	Specific pests die out between plantings of the same crop

Chemical methods

Pesticides include herbicides to kill weeds, fungicides to control fungal diseases, insecticides to kill insect pests, molluscicides to kill mollusc pests and nematicides to kill nematode pests. Herbicides are weedkillers. Selective herbicides have a greater effect on certain plant species (broad-leaved weeds). Selective herbicides work by over-stimulating plant metabolism and killing the leafy part of the plant. The substances are absorbed through green leaves and so kill broad-leaved plants quickly but not narrow-leaved ones such as cereals whose leaves absorb very little of the substance. Systemic herbicides are absorbed and transported through the vascular system of the weed and totally destroy all parts of it, so preventing regrowth.

Fungicides kill fungal parasites of crops and are often sprayed onto crops. Fungicides can be used as a protective measure when environmental conditions and disease forecasts suggest fungal infections are likely.

Systemic insecticides, molluscicides and nematicides spread through the vascular systems of plants and kill pests that feed on the plants. Problems with pesticides include toxicity to non-target species, persistence in the environment, bioaccumulation or biomagnification in food chains and the production of resistant populations of pests. Bioaccumulation is a build-up of a chemical in an organism. Biomagnification is an increase in the concentration of a chemical between trophic levels. One of the most effective early insecticides was DDT. Unfortunately this substance is persistent, which means that, after spraying, it remains in the ecosystem and can be passed along food chains, gradually accumulating in the bodies of predatory animals at the end of the chain.

Figure 3.12 shows how DDT can accumulate along a food chain. Loss of the predators sets the chain out of balance and ultimately the system collapses. This effect is especially damaging when the pesticide leaches in an uncontrolled way into water near to farming areas.

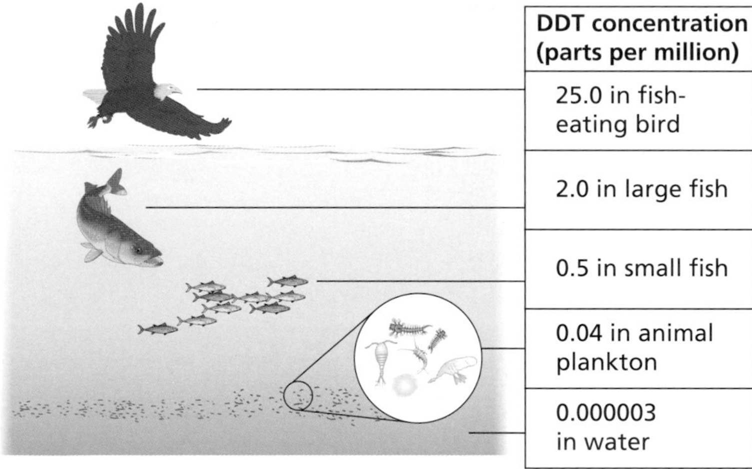

Figure 3.12 Bioaccumulation of DDT along a food chain

Biological control

Weeds, other pests and diseases can also be controlled by biological control and integrated pest management. In biological control, the control agent is a natural predator, parasite or pathogen of the pest. The following table shows examples of the various categories of biological control organisms.

Category	Example of pest controlled
Predator	Ladybirds act as predators of adult greenfly
Parasite	*Encarsia* wasp larvae are parasitic on whiteflies
Pathogen	*Bacillus thuringiensis* causes disease in caterpillars

Biological control has the advantage of not requiring chemicals that could persist in the ecosystem and cause unintended damage. Biological control works especially well in closed systems such as greenhouses, where the control agent cannot escape into the wider environment and cause unintended problems.

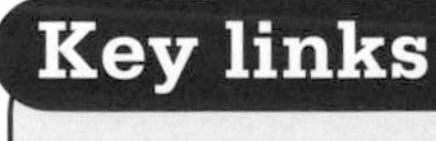

Key links

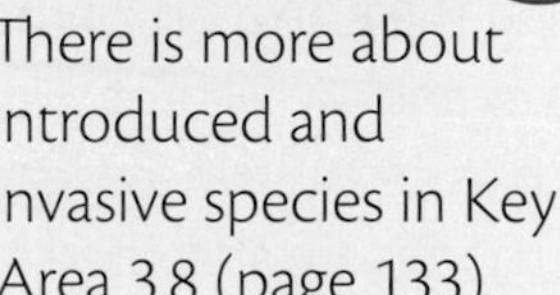

There is more about introduced and invasive species in Key Area 3.8 (page 133).

There are risks linked to the escape of biological control agents into natural ecosystems. If the control agents are introduced from a different part of the world, they may be free from predators, parasites and disease and may become an invasive species, parasitise, prey on or be a pathogen of other species. Their numbers could increase rapidly and they may threaten native species.

Integrated pest management (IPM)

Integrated pest management uses a combination of methods to control pest numbers. It combines cultural, chemical and biological control methods and allows reduction in the application of chemical pesticides.

Animal welfare

Domesticated animals should behave in natural ways, live free from disease and grow vigorously. Methods of livestock production should ensure the well-being of the animals involved.

There are costs, benefits and ethics associated with providing different levels of animal welfare in livestock production.

Intensive farming is less ethical than free-range farming due to poorer animal welfare but is often more cost effective, generating higher profit, as costs are lower. Improving conditions for animals is expensive and results in food that is more costly for human consumers. Free-range farming requires more land and is more labour intensive. However, contented animals grow and reproduce better, and produce higher quality meat, milk and eggs. Free-range products can be sold at a higher price and animals have a better quality of life.

Ethical questions involve evaluating human moral conduct. Poor welfare of domesticated animals could be seen as unethical. Is the need to provide better food security a higher priority than the need to behave ethically? In the UK, the Farm Animal Welfare Council (FAWC) advises the government on the changes to legal regulations needed to ensure animal welfare.

Indicators of poor animal welfare

Poor animal welfare is often indicated by changes to behaviour, leading to reproductive failure. Some examples are shown in the following table.

Behaviour indicating poor welfare	Example
Stereotypy	Repetitive actions such as aimless chewing movements in pigs
Misdirected behaviour	Inappropriate use of normal behaviour, such as over-grooming of feathers by chickens, leading to feather damage
Abnormal activity levels (very low/apathy, or very high/hysteria)	Hyper-aggressive stamping and head lowering in bulls
Failure in sexual or parental behaviour	Rejection or abandoning of offspring by female sheep

Example 

Five freedoms

The welfare of an animal includes its physical and mental state. Good animal welfare implies both fitness and a sense of well-being. The five freedoms are five aspects of animal welfare that are under human control. The five freedoms have been adopted by professional groups, such as vets, and animal welfare organisations, such as the World Organisation for Animal Health, the RSPCA and the American Society for the Prevention of Cruelty to Animals.

The five freedoms as currently expressed are:

1. Freedom from hunger or thirst by ready access to fresh water and a diet to maintain full health and vigour
2. Freedom from discomfort by providing an appropriate environment including shelter and a comfortable resting area
3. Freedom from pain, injury or disease by prevention or rapid diagnosis and treatment
4. Freedom to express normal behaviour by providing sufficient space, proper facilities and company of the animal's own kind
5. Freedom from fear and distress by ensuring conditions and treatment which avoid mental suffering

Key words

Animal welfare – relating to activities that are designed to be humane to livestock while maximising their yield

Annual weed – a weed that completes its life cycle in one year, which has rapid growth, high seed output and long-term seed viability

Bioaccumulation – build-up of a chemical in an organism

Biomagnification – increase in the concentration of a chemical between trophic levels

Biological control – method of controlling pests using natural predators, parasites or pathogen of the pest

Crop plant pests – organisms such as insects, nematode worms and molluscs which reduce the yield of crops

Cultural – method of crop protection based on human behaviours and activities such as ploughing, weeding and crop rotation

Disease forecasting – management system used to predict the occurrence of plant diseases; applications of fungicide based on disease forecasts are more effective than treating diseased crops

Fungicide – chemical substance which kills fungal pest species

Integrated pest management (IPM) – a combination of chemical, biological and cultural means to control pests

Misdirected behaviour – inappropriate use of normal behaviour, such as over-grooming of feathers by chickens

Perennial weed – weed plant with storage organs and vegetative reproduction which persists in the community by continuing to grow year after year

Persistent – unable to be broken down by enzymes

Pesticides – these include herbicides to kill weeds, fungicides to control fungal diseases, insecticides to kill insect pests, molluscicides to kill mollusc pests and nematicides to kill nematode pests

Selective herbicides – weedkillers that have a greater and targeted effect on certain plant species (broad-leaved weeds)

Stereotypy – repetitive actions such as aimless chewing movements in pigs

Systemic herbicide – weedkiller that spreads through the vascular system (phloem) of the plant and prevents regrowth

Vegetative reproduction – a form of asexual reproduction in plants

Questions ?

Short answer (1 or 2 marks)

1 Name **two** groups of organisms that are carried by invertebrates and which are often responsible for causing plant diseases. (2)

2 Give **two** properties of an annual weed. (2)

3 Give **one** property of a perennial weed. (1)

4 Name the type of herbicide that has a greater effect on broad-leaved weeds. (1)

5 Explain why systemic herbicides are effective against weeds with underground storage organs. (2)

6 Explain the meaning of the term *integrated pest management* (IPM). (1)

7 Give **two** examples of behaviour which could indicate poor animal welfare. (2)

Longer answer (3–10 marks)

8 Describe the adaptations of annual and perennial plants which allow them to be successful weeds of crop plants. (4)

9 Give an account of biological control and outline a risk linked to use of this method of crop protection. (4)

10 Write notes on the costs, benefits and ethics of providing different levels of animal welfare in livestock production. (5)

11 Give an account of chemical methods used to protect plants from named pests and the environmental damage that can result from their use. (9)

Answers are on page 137.

Key Areas 3.5 and 3.6

Symbiosis and social behaviour

Key points

1 **Symbiosis** is a coevolved intimate relationship between members of two different species. ☐

2 Types of symbiotic relationship include parasitism and **mutualism**. ☐

3 A **parasite** benefits in terms of gain in energy or nutrients, whereas its **host** is harmed by the loss of these resources. ☐

4 Parasites often have limited metabolism and cannot survive without a host. ☐

5 Parasites can be transmitted to new hosts using direct contact, **resistant stages** and **vectors**. ☐

6 Some parasitic life cycles involve **intermediate (secondary) hosts** to allow them to complete their life cycle. ☐

7 Mutualism is a form of symbiosis in which both mutualistic partner species benefit in an interdependent relationship. ☐

8 Many animals live in social groups and have behaviours that are adapted to group living, such as **social hierarchy**, **cooperative hunting** and social defence. ☐

9 Social hierarchy is a rank order of individuals within a group of animals consisting of **dominant** and **subordinate** members. ☐

10 In a social hierarchy, dominant individuals carry out **ritualistic threat displays** while subordinate animals carry out **appeasement behaviour** to reduce conflict. ☐

11 Social hierarchies increase the chance of the dominant animal's favourable alleles being passed on to offspring. ☐

12 Animals often form **alliances** in social hierarchies to increase their social status within the group. ☐

13 Cooperative hunting may benefit subordinate animals as well as dominant ones, as they may gain more food than by foraging alone and less energy is used per individual. ☐

14 Cooperative hunting enables larger prey to be caught and increases the chance of success. ☐

15 Social defence strategies increase the chance of survival as some individuals can watch for predators while others can forage for food. ☐

16 Some groups adopt specialised formations when under attack to protect their young. ☐

17 An **altruistic behaviour** harms the **donor** individual but benefits the **recipient**. ☐

18 Behaviour that appears to be altruistic can be common between a donor and a recipient if they are related (kin). ☐

18 The donor will benefit in **kin selection** in terms of the increased chances of survival of shared genes in the recipient's offspring or future offspring. ☐

20 **Reciprocal altruism**, in which the roles of the donor and the recipient are later reversed, often occurs in social animals. ☐

21 **Social insects** include bees, wasps, ants and termites. ☐

⇨

⇨

22 Social insects have a society in which only some individuals (queens and drones) contribute reproductively. ☐
23 Most members of the colony are sterile workers who cooperate with and raise close relatives. ☐
24 Other examples of workers' roles include defending the hive, collecting pollen and carrying out waggle dances to show the direction of food. ☐
25 Sterile workers raise relatives to increase the survival of shared genes. This is an example of kin selection. ☐
26 **Primates** have a long period of **parental care** to allow learning of complex social behaviour. ☐
27 Complex social behaviours support the social hierarchy. ☐
28 This reduces conflict through ritualistic display and appeasement behaviour. ☐
29 Examples of appeasement behaviour include grooming, facial expression, body posture and sexual presentation. ☐
30 Alliances which form between individuals, are often used to increase social status within the group. ☐

Summary notes

Symbiosis

Symbiosis is an intimate ecological relationship between members of two different species that have coevolved alongside each other over millions of years. Types of symbiotic relationship include parasitism and mutualism.

Parasitism

A parasite has an intimate ecological relationship with its host. The parasite species is dependent on its host species and benefits in terms of gain in energy. Whereas the host is harmed by the loss of these resources. Parasites often have limited metabolism, which prevents their survival out of contact with their host.

Parasites are transmitted to their hosts by direct contact, through resistant stages or by vectors. Some parasites have life cycles that involve an intermediate (secondary) host. The following table shows some examples of parasite species.

Parasite species	Host	Intermediate (secondary) host	Transmission
Human head lice	Human	None	Direct physical contact between humans
Cat fleas	Domestic cat	None	Direct physical contact between cats when adult fleas passed, or by resistant-stage larvae or pupae of the flea being picked up from the environment
Malaria parasite (*Plasmodium*)	Human	None	Mosquito acts as a vector, which carries the parasite stages from human to human through biting
Pork tapeworm	Human	Pig (also damaged by stages of the tapeworm)	Consumption of undercooked pork from an infected pig by a human

Figure 3.13 shows the life cycle of the cat flea.

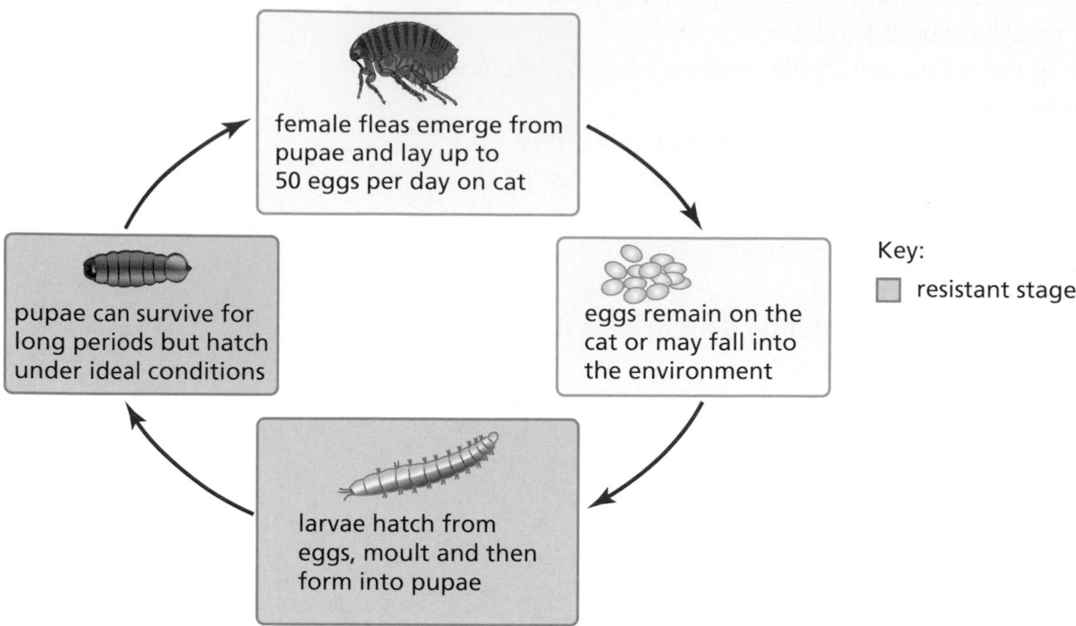

Figure 3.13 Life cycle of the cat flea

Figure 3.14 shows how an intermediate (secondary) host is involved in the life cycle of the pork tapeworm.

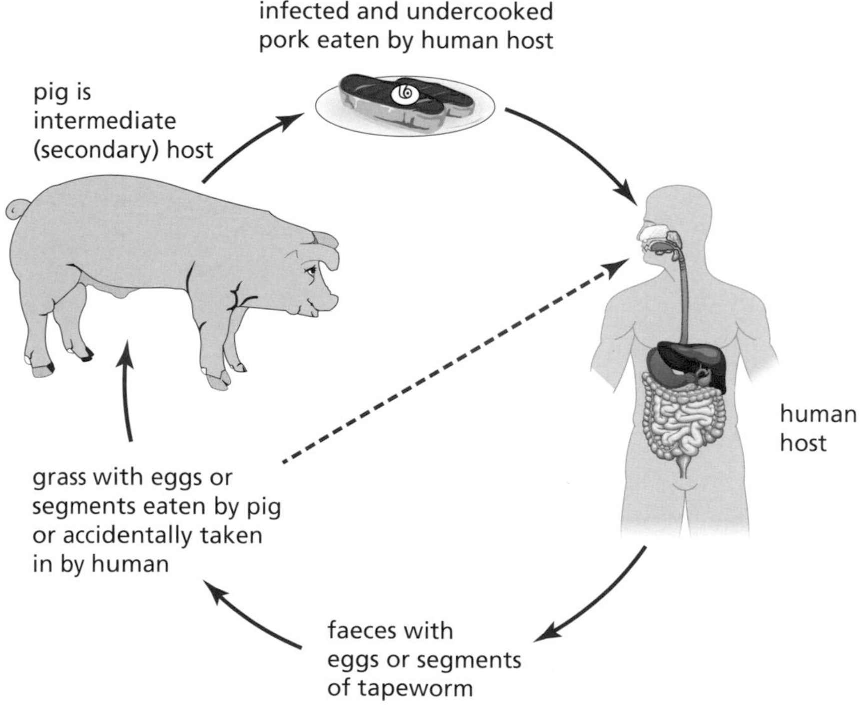

Figure 3.14 Pork tapeworm life cycle

Mutualism

Mutualism is a form of symbiosis in which both mutualistic partner species benefit in an interdependent relationship. The following table shows some examples of mutualistic relationships between species.

Partner species	Notes on interdependence
Herbivorous grazers (e.g. cattle) with cellulose-digesting bacteria	Their gut provides a safe, warm, moist habitat with a continuous supply of food for the cellulose-digesting bacteria that live there. The bacteria digest cellulose to produce simple sugars used by the herbivores as an energy source because many do not have the required digestive enzymes themselves.
Polyp stages of coral with photosynthetic *Zooxanthella* algae	The soft-bodied part of sessile corals provides a safe, nitrogen-rich habitat for *Zooxanthella* algae, which live there and produce photosynthetic sugars that supply carbohydrate to the coral polyp, which it would otherwise lack.

Social behaviour

Many animals live in social groups and have behaviours that are adapted to group living, such as social hierarchy, cooperative hunting and social defence.

Social hierarchy is a rank order of individuals within a group of animals consisting of dominant and subordinate members. In a social hierarchy, dominant individuals carry out ritualistic (threat) displays while subordinate animals carry out appeasement behaviour to reduce conflict. Social hierarchies increase the chance of the dominant animal's favourable alleles being passed on to offspring. Animals often form alliances in social hierarchies to increase their social status within the group.

Cooperative hunting may benefit subordinate animals as well as dominant ones, as they may gain more food than by foraging alone and less energy is used per individual. Cooperative hunting enables larger prey to be caught and increases the chance of success.

Social defence strategies increase the chance of survival as some individuals can watch for predators while others can forage for food. Some groups adopt specialised formations when under attack to protect their young. Some behavioural adaptations are summarised in the following table.

Behavioural adaptation	Species as an example	Survival value
Social hierarchy	Grey wolf	Lowers aggression and saves energy Experienced leadership guaranteed Most favourable alleles passed on
Cooperative hunting	African wild dogs	Larger prey can be killed Subordinate animals benefit Less energy used per individual
Social defence	Baboons	Early warning can be given Younger individuals defended Predators intimidated or confused

Altruistic behaviour

Altruistic behaviour involves a donor harming itself to the benefit of the recipient. Behaviour that appears to be altruistic can be common between a donor and a recipient if they are related (kin). The donor will benefit in kin selection in terms of the increased chances of survival of shared genes in the recipient's offspring or future offspring.

Reciprocal altruism, in which the roles of the donor and the recipient are later reversed, often occurs in social animals. Explanations for different types of altruism are shown in the following table.

Altruism	Example	Explanation
Reciprocal	Vampire bats who have hunted successfully might share food at the roost with those who have not	The successful hunter on one occasion might be unsuccessful later and need to obtain food from a previous recipient
Kin selection	Donor long-tailed tits with no offspring might feed the recipient offspring of other parents in times of food shortage	Long-tailed tits live in loose colonies with related individuals, so that recipient offspring of one parent might share some of the donor's genes

Social insects

Some insects live in social colonies, for example bees, wasps, ants and termites. Social insects have a society in which only some individuals (queens and drones, Figure 3.15) contribute reproductively. Most members of the colony are sterile workers who cooperate with and raise close relatives. Other examples of workers' roles include defending the hive, collecting pollen and carrying out waggle dances to show the direction of food. Sterile workers raise relatives to increase the survival of shared genes. This breeding system is an example of kin selection.

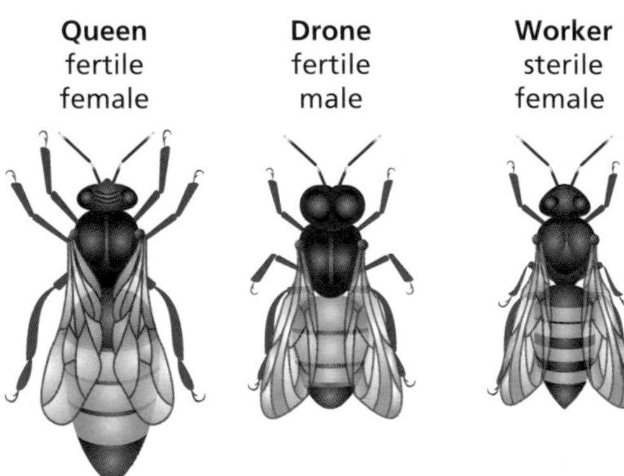

Figure 3.15 Honey bees. The queen bee (a fertile female) can be fertilised by a male drone

Primate behaviour

Primates form the mammal group that includes monkeys, apes and humans. Primate offspring are born in a generally helpless state and have a long period of parental care. This allows time for the learning of complex social behaviours that support the social hierarchy. These behaviours include some designed to reduce conflict within the social group, such as ritualistic, display and appeasement behaviours.

The following table shows some behaviours that are important in chimpanzees.

Chimpanzee social group behaviour	Description	Function
Grooming	Includes the preening of one animal's coat by another	Reduction of tension and strengthening of alliances to increase social status in the group; strengthening of bonds between individuals
Facial expressions	Include eye closing, teeth baring, mouth opening	Act as signals to indicate position in dominance hierarchy and avoid conflict
Body postures	Include lowering of body position and bowing actions	Act as signals to emphasise position in dominance hierarchy and avoid conflict
Sexual presentation	Includes the presentation of genitalia by females to males	Acts as a signal by females to appease dominant males and avoid aggression

Alliances form between individuals and are often used to increase social status within the group. Benefits of increased social status include protection and access to better food. Females with higher social status have higher infant survival than their low-ranking counterparts.

Key words

Alliance – link between individuals in primate social groups, often used to increase social status within the group
Altruistic behaviour – behaviour which harms a donor but benefits the recipient
Appeasement behaviour – behaviour carried out by subordinate animals to appease a dominant individual and reduce conflict
Cooperative hunting – hunting behaviour in which individuals work together to catch prey
Dominant – animal ranked at the top of a social hierarchy
Donor – organism that carries out the altruistic behaviour to benefit the recipient
Host – organism that is harmed by the loss of energy and nutrients to a parasite
Intermediate (secondary) host – organism involved in the life cycle of a parasite but that is separate from the main host
Kin selection – situation in which the donor benefits in terms of the increased chance of survival of shared genes in the recipient's offspring or future offspring
Mutualism – symbiosis in which both partners benefit from the relationship
Parasite – organism in the symbiotic relationship that benefits in terms of energy or nutrients with the host being harmed by the loss of these resources
Parental care – activities performed by parents that increase the survival chances of their young
Primates – mammalian group that includes monkeys, apes and humans
Recipient – organism that benefits from the altruistic behaviour of the donor
Reciprocal altruism – a selfless behaviour which is returned by the original recipient to the original donor
Resistant stage – some parasites use resistant larvae and pupae, which can survive adverse environmental conditions until a new host comes in contact with them
Ritualistic threat display – behaviour such as body posture, raised hackles and baring teeth, used instead of physical aggression
Social hierarchies – a rank order within a group of animals consisting of a dominant and subordinate members
Social insect – insect which lives in a complex social colony, such as bees, wasps and termites
Subordinate – an animal ranked below the dominant individual
Symbiosis – an intimate coevolved relationship between two different species
Vector – an organism that carries and transmits the parasite into a new host organism

Questions ?

Short answer (1 or 2 marks)

1 Describe the features of a symbiotic relationship. (2)

2 Give **two** ways in which parasites can be transmitted to new hosts. (2)

3 Describe the relationship between a parasite and its host. (2)

4 Grey wolves live in packs in which a social hierarchy exists. The animals use cooperative hunting techniques.

a) Give a definition of the term *dominance hierarchy*. (1)

b) State **two** advantages to wolves of using cooperative hunting. (2)

5 In honey bees, worker individuals are sterile but function to ensure that offspring of their relatives are fed.

a) Explain the altruistic behaviour of the worker bees. (2)

b) Give the term that describes altruistic behaviour towards relatives. (1)

6 Explain the benefit to primates in forming alliances between individuals within a group. (2)

Longer answer (3–10 marks)

7 Give an account of symbiosis under the following headings:

a) parasitism (5)

b) mutualism. (2)

(total = 7)

8 Give an account of behaviour in social insects. (5)

9 Give an account of primate behaviour under the following headings:

a) parental care (2)

b) behaviours for the reduction of conflict. (2)

(total = 4)

Answers are on page 138.

Key Area 3.7

Components of biodiversity

Key points 

1 Components of **biodiversity** are **genetic diversity**, **species diversity** and **ecosystem diversity**. ☐
2 Genetic diversity is the number and frequency of all the alleles within a population. ☐
3 If one population of a species dies out then the species may have lost some of its genetic diversity, and this may limit its ability to adapt to changing environmental conditions. ☐
4 Species diversity comprises the number of different species in an ecosystem (the **species richness**) and the proportion of each species in the ecosystem (the **relative abundance**). ☐
5 A community with a **dominant species** has lower species diversity than one with the same species richness but no particularly dominant species. ☐
6 Ecosystem diversity refers to the number of distinct ecosystems within a defined area. ☐

Summary notes

Measuring biodiversity

Biodiversity can be measured using components such as genetic, species and ecosystem diversity, as detailed in the following table.

Diversity component	Definition
Genetic	The number and frequency of alleles in a population
Species	The number of different species in an ecosystem (the species richness) and the proportion of each species in the ecosystem (the relative abundance)
Ecosystem	The number of distinct ecosystems within a defined area

Dominant species

Some ecosystems are characterised by dominant species. A Scottish heather moor, for example, is dominated by ling heather (*Calluna vulgaris*). This reduces the species diversity in the ecosystem because the ling heather is so abundant that it takes most resources from the habitat and contributes the most to productivity. The relative abundance of other plant species is much lower.

Key words

Biodiversity – variety and relative abundance of species
Dominant species – most abundant species in an ecosystem
Ecosystem diversity – the number of distinct ecosystems within a defined area
Genetic diversity – the number and frequency of all the alleles within a population
Relative abundance – the numbers of a species compared with others in a community
Species diversity – the number of different species in an ecosystem (the species richness) and the proportion of each species in the ecosystem (the relative abundance)
Species richness – the number of different species in an ecosystem

Questions

Short answer (1 or 2 marks)

1 Give the meaning of the term *ecosystem diversity*. (1)
2 Give **one** component of genetic diversity. (1)
3 Give the meaning of the term *species richness*. (1)
4 Describe the effect of a dominant species on the species diversity of a community. (1)

Longer answer (3–10 marks)

5 Give an account of biodiversity and its measurement with reference to genetic, species and ecosystem diversity. (5)

Answers are on page 139.

Key Area 3.8

Threats to biodiversity

Key points

1 With over-exploitation, populations can be reduced to a low level but may still recover. ☐
2 Some species have a naturally low genetic diversity in their population and yet remain viable. ☐
3 The **bottleneck effect** relates to small populations that may lose the genetic variation necessary to enable evolutionary responses to environmental change. ☐
4 For many species, the loss of genetic diversity in small populations can be critical, as inbreeding can result in poor reproductive rates. ☐
5 The clearing of habitats has led to habitat fragmentation. ☐
6 Degradation of the edges of **habitat fragments** results in increased competition between species as the fragment becomes smaller. This may result in a decrease in biodiversity. ☐
7 More isolated fragments and smaller fragments exhibit lower species diversity. ☐
8 To remedy widespread habitat fragmentation, isolated fragments can be linked with **habitat corridors**. ☐
9 The habitat corridors allow movement of animals between fragments, increasing access to food and choice of a mate. ☐
10 The formation of habitat corridors may lead to recolonisation of small fragments after local extinctions. ☐
11 **Introduced (non-native) species** are those that humans have moved either intentionally or accidentally to new geographic locations. ☐
12 Those species that become established within wild communities are termed **naturalised species**. ☐
13 **Invasive species** are naturalised species that spread rapidly and eliminate native species, therefore reducing species diversity. ☐
14 Invasive species may well be free of the predators, parasites, pathogens and competitors that limit their population in their native habitat. ☐
15 Invasive species may prey on **native species**, out-compete them for resources or hybridise with them. ☐

Summary notes

Exploitation

Humans exploit natural resources for food, raw materials and space. Over-exploitation, however, involves resources being consumed at a rate greater than they can be replaced. If over-exploitation is halted soon enough, populations of over-exploited species can be reduced to a low level but may still recover. An example of an over-exploited species and its potential recovery is provided by the fishing of cod. Exploitation of cod turned into over-exploitation when over-fishing caused depletion of stock. Quotas have been introduced by governments in recent years and there are some signs that cod stocks might be recovering – time will tell.

The bottleneck effect

Some populations can be reduced drastically. If the surviving population is very small, it might have lost most of its genetic variability. If the survivors are genetically similar, their inbreeding can then lead to poor reproductive rates, inbreeding depression and to further loss of genetic diversity, which could mean that the species might be unable to adapt to environmental change in the future. This is known as the bottleneck effect. For example, the northern elephant seal, whose population was drastically reduced by over-hunting in the nineteenth century, has recovered in recent years although the genetic diversity of the modern population is very low. Some species have a naturally low genetic diversity in their population and yet remain viable.

Key links

There is more about inbreeding depression in Key Area 3.2 (page 111).

Habitat fragmentation

Habitat fragmentation often occurs when humans take over an ecosystem. Forests have been cleared for agriculture and housing and to use the timber that they yield. This practice leaves behind remnants of the original habitat known as habitat fragments. Degradation of the edges of habitat fragments results in increased competition between species as the fragment becomes smaller. Edge species that live at the edges of habitat fragments often colonise the centres of smaller fragments and can cause decline in the numbers of other species, reducing overall biodiversity. More isolated fragments and smaller fragments exhibit a lower species diversity. The collective habitat fragments support lower species richness and abundance than the original habitat.

It has been suggested that the creation of corridors of habitat between the small fragments (Figure 3.16) might allow species to recolonise them following local extinction. The species are able to move, feed and even mate along the corridors. This principle has been applied to hedgerows linking forest fragments on agricultural land and the creation of motorway underpasses for wildlife. However, these might not increase biodiversity because they do not provide for species that require continuous habitat with no breaks. Also, there is the suggestion that they might have a negative effect because they could allow the spread of disease between fragments.

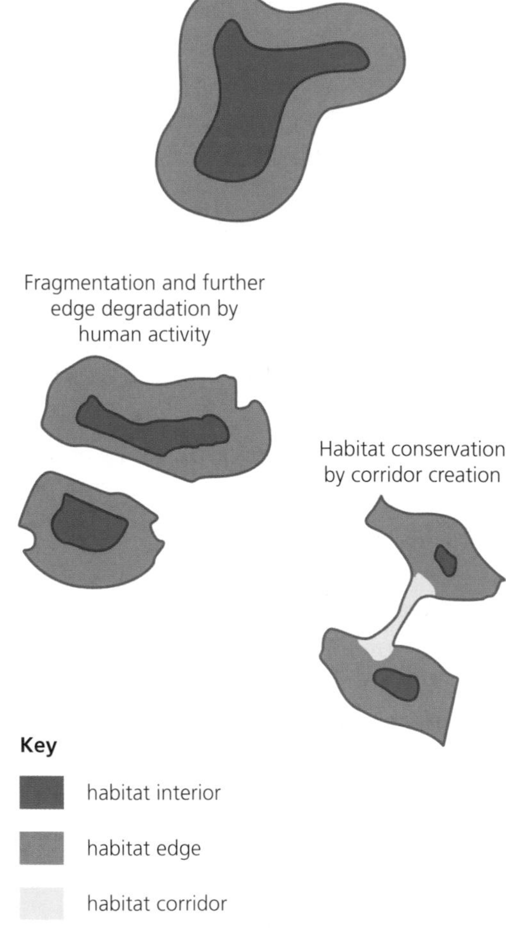

Figure 3.16 Habitat fragmentation favours edge species over interior species. Corridor creation allows movement of species and the recolonisation of small fragments

Introduced, naturalised and invasive species

Introduced species are those that are non-native and have been moved by humans either intentionally or accidentally to new geographical locations. Some might become naturalised, which means that they become established within wild communities in their new location. Naturalised species can spread rapidly because they are free from the natural predators, parasites, pathogens or competitors that limited their population in their original habitat. They can become invasive and threaten native species by preying on them, out-competing them for resources and, in some cases, by hybridising with them.

Example 

A good example of problems caused by an introduced animal becoming invasive is the cane toad, which was introduced to Australia in the 1930s as a biological control for beetles, which were damaging sugar cane crops. The toads were poisonous to predators and ate a variety of native species as well as beetles, including marsupial mammals. The toads are still increasing and threatening native biodiversity as invasive pests.

Key words

Bottleneck effect – inability of a species to evolve due to lack of genetic diversity
Habitat corridor – link that allows movement of animals between habitat islands or fragments, increasing access to food and choice of a mate
Habitat fragment – very small area of isolated habitat
Introduced (non-native) species – species that humans have moved either intentionally or accidentally to a new geographic location
Invasive species – naturalised species that spreads rapidly and eliminates native species, therefore reducing species diversity
Native species – (indigenous) species occurring naturally in its ecosystem, having evolved there
Naturalised species – species that becomes established within a wild community

Questions 

Short answer (1 or 2 marks)

1 Describe what is meant by the bottleneck effect. (2)
2 Explain why the loss of genetic diversity in small populations can be critical. (2)
3 Describe the impact resulting from the degradation of the edges of habitat fragments. (2)
4 Explain how a habitat corridor can increase biodiversity after local extinction. (2)
5 State what is meant by the following terms:
- **a)** introduced species (1)
- **b)** naturalised species (1)
- **c)** invasive species. (2)

Longer answer (3–10 marks)

6 Give an account of habitat fragmentation and the measures that can be taken to minimise its effects. (5)
7 Give an account of invasive species in an ecosystem. (5)

Answers are on page 139.

Answers

Key Area 3.1a

Short answer questions

1 population increase [1]

2 a) production of food by green plants using energy from light [1]

b) an organism's position in a food chain [1]

3 plant a greater area of crop; use fertilisers; use pesticide; use biological control methods; grow improved/GM strains of crop plants (other answers possible) [any 3 = 2, 2/1 = 1]

4 crops are producers and livestock are consumers; energy is lost between each trophic level of a food chain [1 each = 2]

Longer answer questions

5 a) sufficient quantities of food are produced; sufficient quality of food is produced; ability to distribute/spread food through the population [1 each = 3]

b) knowledge required to use food properly; ability to guarantee food security; over longer periods [1 each = 2]

[total = 5]

Answers

Key Area 3.1b

Short answer questions

1 absorbed; reflected; transmitted [any 2 = 2]

2 absorption spectrum shows which wavelengths of light are absorbed by an isolated pigment; action spectrum shows which wavelengths of light cause most photosynthesis to occur [1 each = 2]

3 carotenoid pigments extend the wavelengths of light absorbed by photosynthesis; and pass the energy trapped onto chlorophyll [1 each = 2]

4 photolysis (of water) [1]

5 release energy; to generate ATP by ATP synthase [2]

6 RuBisCo [1]

7 NADP/NADPH [1]

8 converted into glucose; or used to regenerate RuBP [2]

Longer answer questions

9 pigments absorb light energy; main pigment is chlorophyll a; carotenoids extend the range of wavelengths absorbed; light energy absorbed excites electrons; electrons pass through the electron transport chain; ATP synthase produces/synthesises ATP; some light energy absorbed by pigments is used for photolysis/to split a molecule of water into hydrogen and oxygen; ATP and NADPH are produced for the carbon fixation stage [any 6 = 6]

10 paper chromatography/thin layer chromatography; extraction of leaf pigment; spotting/loading of pigment onto paper/slide; use of solvent to separate pigments sample [any 3 = 3]

11 a) light energy excites electrons in pigment molecules; high-energy electrons pass through an electron transport chain; releasing energy to generate ATP; by ATP synthase; energy also used to split water into oxygen, which is released; and hydrogen, which is transferred to the coenzyme NADP [any 5 = 5]

b) carbon dioxide joined to RuBP by RuBisCo; to produce 3-phosphoglycerate; ATP used to phosphorylate 3-phosphoglycerate to form G3P; hydrogen from NADPH used to form G3P; G3P forms glucose; some G3P regenerates RuBP [any 4 = 4]

[total = 9]

Answers

Key Area 3.2

Short answer questions

1 higher yield; higher nutritional value; resistance to pests/diseases/herbicides; characteristics suited to rearing/harvesting; characteristics suitable for survival in particular habitats/environmental conditions [any 2 = 2]

2 the process of crossing organisms of similar genotypes/related individuals; to increase homozygosity of desired/useful characteristic/allele OR until the population breeds true due to the elimination of heterozygotes [1 each = 2]

3 the breeding together of genetically similar organisms/related individuals; resulting in an increase in the frequency of recessive deleterious/harmful alleles [1 each = 2]

4 to produce a new F_1 cross-breed population/individuals with improved characteristics/yield/hybrid vigour/increased disease resistance/increased growth rate [1]

5 the F_2 offspring show too much variation/have a lower percentage showing the improved characteristic/hybrid vigour [1]

6 genomic sequencing (1)

Longer answer questions

7 single genes for desirable characteristics can be inserted into the genomes of crop plants; creating genetically modified plants with improved characteristics; this includes insertion of the Bt toxin gene into plants for pest resistance; this protects crops from being eaten/damaged by pests; and less pesticide needs to be used; the insertion of the glyphosate resistance gene for herbicide tolerance; allows other weeds/competing plants to be removed; increasing crop yield [any 6]

[total = 6]

Answers

Key Areas 3.3 and 3.4

Short answer questions

1 bacteria/virus/fungi [any 2 = 2]

2 rapid growth/short life cycle/high seed output/long-term seed viability [any 2 = 2]

3 storage organs/vegetative reproduction [any 1 = 1]

4 selective herbicide [1]

5 spread through vascular system; prevent regrowth [2]

6 combination of chemical, biological and cultural methods used [1]

7 stereotypy/misdirected behaviour/failure in sexual OR parental behaviour/altered levels of activity OR apathy OR hysteria [any 2 = 2]

Longer answer questions

8 annual weeds have a rapid growth; short life cycle; high seed output; long-term seed viability. Perennial weeds have storage organs; vegetative reproduction [any 4 = 4]

9 use of natural predator/parasite/pathogen to control pest numbers; better used in enclosed situation/greenhouse; control organism may be free of predators/parasites/pathogen; could become invasive/a threat to native species; parasitise/prey on OR be a pathogen of other species [any 4 = 4]

10 free range requires more land; free range is more labour intensive; free range produce can be sold at a higher price; animals have a better quality of life; intensive farming often creates conditions of poor animal welfare; intensive farming is often more cost effective/generates higher profit/has lower costs; intensive farming is less ethical than free range farming due to poorer animal welfare [any 5 = 5]

11 pests can multiply rapidly in agricultural communities; any two invertebrate examples from insects/molluscs/nematode worms; any two from weeds/fungi/viruses/bacteria; pesticide; any two examples from herbicide/insecticide/fungicide; can be selective or systemic; fungicide can be applied based on fungal disease forecasts [any 4 = 4]
might be toxic to animals; might persist in the environment; might accumulate/bioaccumulation/be magnified in food chains/biomagnification; might lead to damage/imbalance in natural populations; might produce selection pressure on a population; might result in resistant populations [any 5 = 5]
[total = 9]

Answers

Key Areas 3.5 and 3.6

Short answer questions

1 coevolved AND intimate relationship; between members of two different species [2]

2 direct contact; resistant stages; vectors [any 2 = 2]

3 parasite benefits in terms of energy or nutrients; host is harmed by the loss of these resources [2]

4 a) dominance hierarchy is a rank order/pecking order of individuals in a social grouping of animals [1]
b) allows larger kills to be made; increases success rate of hunts; energy gained in food greater than that lost in hunting [any 2 = 2]

5 a) worker bees feed the offspring of relatives because they have shared genes; the feeding helps ensure that the offspring survive [1 each = 2]
b) kin selection [1]

6 access to better food; females with a higher social status have higher infant survival than their low-ranking counterparts; protection [any 2 = 2]

Longer answer questions

7 a) parasites gain energy/nutrients; host is harmed by loss of these resources; parasites have more limited metabolism; cannot survive out of contact with host; are transmitted by direct contact; resistant stages or vectors; some have secondary hosts [any 5 = 5]
b) mutualism benefits both partners; partners are interdependent; example – bacteria in herbivore gut/photosynthetic algae in coral polyps (other answers possible) [any 2 = 2]
[total = 7]

8 social insects include bees/wasps/ants/termites; only some individuals/queens and drones contribute reproductively; sterile workers cooperate with close relatives to raise relatives; other examples of workers' roles include defending the hive/collecting pollen/carrying out waggle dances to show the direction of food; sterile workers raise relatives to increase the survival of shared genes; this is an example of kin selection [any 5 = 5]

9 a) long period of parental care in primates; allows opportunity to learn complex social behaviours [1 each = 2]
b) ritualistic display or specific example; appeasement behaviour OR specific example; general examples – grooming/facial expression/body posture/sexual presentation [any 2 = 2]
[total = 4]

Answers

Key Area 3.7

Short answer questions

1 the number of distinct ecosystems within a defined area [1]

2 the number of all the alleles within a population; the frequency of all the alleles within a population [any 1 = 1]

3 the number of different species in an ecosystem [1]

4 it has a lower species diversity [1]

Longer answer questions

5 genetic diversity is the number of alleles in a population; and frequencies of alleles in a population; loss of a population can result in loss of genetic diversity of a species [any 2 = 2]

number of different species is species richness; relative abundance is the proportion of each species present [1 each = 2]

ecosystem diversity is the number of distinct ecosystems in a defined area [1]

[total = 5]

Answers

Key Area 3.8

Short answer questions

1 when a small population loses the genetic variation; necessary to enable evolutionary responses to environmental change [2]

2 loss of the genetic variation necessary to enable evolutionary responses to environmental change; poor reproductive success/rates [2]

3 increased competition between species as the fragment becomes smaller; decrease in biodiversity [2]

4 allows movement of animals between fragments; increasing/giving access to food/choice of mate; this may lead to recolonisation of small fragments (after local extinctions) [any 2 = 2]

5 a) introduced species have been moved intentionally from one geographic location to another by humans [1]

b) naturalised species have become established in wild populations in natural communities [1]

c) invasive species are naturalised, have spread rapidly and eliminated native species; they may be free of natural predators/parasites/pathogens OR they may prey on native species OR may out-compete them [any 2 = 2]

Longer answer questions

6 habitat fragmentation occurs when habitat is split into small parts; fragments support lower species richness than a large area of the same habitat; fragments may be degraded at their edges; increased competition between species in smaller fragments [any 3 = 3]

isolated fragments can be connected by habitat corridors; species can feed/mate within corridors; recolonisation of small fragments after local extinctions [any 2 = 2]

[total = 5]

7 introduced species OR species moved intentionally/accidentally by humans; become naturalised species when they are established (in the new area); spread rapidly; may eliminate/kill off/destroy native/original species; prey on/out-compete/hybridise with native/original species; natural/original predators/parasites/pathogens/competitors are not present in new ecosystem [any 5 = 5]

Practice course assessment: Area 3 (50 marks)

Paper 1 (10 marks)

1 The following absorption spectra were obtained from four different plant extracts using a spectroscope. Black areas indicate light which has been absorbed by the extracts.

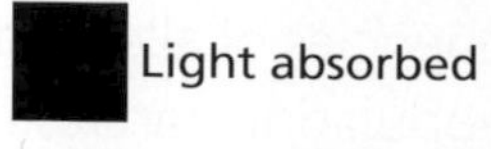

A 

B 

C 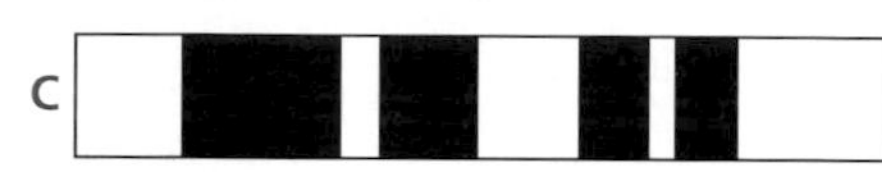

D 

blue green yellow orange red

Which extract contains chlorophyll?

2 The following statements refer to photosynthesis:

1 The enzyme RuBisCo fixes CO_2.

2 Oxygen is released as a by-product.

3 G3P is used for the synthesis of glucose.

Which of the statements also refer correctly to the carbon fixation stage (Calvin cycle)?

A 1 and 2 only

B 1 and 3 only

C 2 and 3 only

D 1, 2 and 3

3 The graph below shows the effect of adding different levels of fertiliser on the yield of a crop plant.

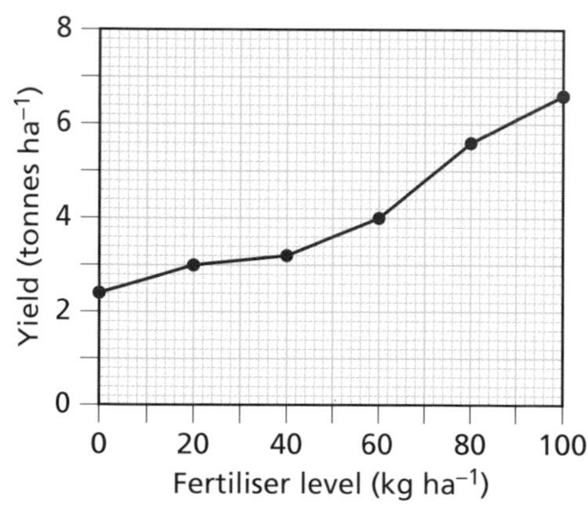

The percentage increase in yield obtained when the fertiliser level is increased from 40 to 80 kg ha^{-1} is

A 24

B 40

C 58

D 75

⇨

4 A farmer used a pesticide to treat an infestation of greenfly that was damaging crops. The concentration of pesticide in the tissues of the greenfly and of the birds that ate them was measured and found to be higher in the birds which ate the greenfly than the greenfly themselves. This example shows that the pesticides can

A biomagnify
B act selectively
C bioaccumulate
D produce resistant organisms.

5 Which of the following is said to occur when an interaction benefits both species in a symbiotic relationship?

A parasitism
B mutualism
C predation
D altruism

6 Adult pork tapeworms live in the intestine of humans. Segments of the adult worm are released in the faeces. The tapeworm embryos that eventually develop from the segments might be eaten by pigs and develop further in their muscle tissue.
Which row in the table below correctly identifies the various roles in the tapeworm life cycle?

	Role of human	Role of embryo	Role of pigs
A	Host	Resistant stage	Secondary host
B	Host	Vector	Secondary host
C	Secondary host	Vector	Host
D	Secondary host	Resistant stage	Vector

7 Altruistic behaviour

A benefits both the donor and the recipient
B benefits the donor and harms the recipient
C harms the donor and benefits the recipient
D harms both the donor and the recipient.

8 Which of the following statements refer to advantages gained by cooperative hunting behaviour?

1 Individuals gain more energy than from hunting alone.
2 Dominant and subordinate animals both benefit.
3 Larger prey can be killed than by hunting alone.

A 1 and 2 only
B 1 and 3 only
C 2 and 3 only
D 1, 2 and 3

9 The species diversity in an ecosystem comprises

A both the genetic diversity and the ecosystem diversity
B species richness and relative abundance
C number and frequency of alleles
D number of distinct ecosystems in a defined area.

10 Over-hunting of the northern elephant seal was responsible for their dramatic decrease in numbers, which resulted in them having very low genetic variation. Present-day populations have all descended from the small number that survived.
What term is used to refer to the loss of genetic variation associated with a serious decline in population?

A over-exploitation
B stabilising selection
C bottleneck effect
D directional selection

⇨

Paper 2 (40 marks)

1 As a result of the increase in human population and concern for food security there is a continuing demand for increased food production.
The table below shows the results of a field trial undertaken to investigate the effect of a fertiliser on the yield of grain obtained per year from an experimental plot.

Crop	Yield of grain (kg per hectare per year)	
	Control	Addition of NPK fertiliser
Maize	260	3200
Wheat	400	2600

a) Calculate the expected increase in the yield of wheat that would be achieved by the addition of NPK fertiliser in a 50-hectare field over a five-year period. (1)

b) Plant field trials are often carried out in a range of environments to compare the performance of different cultivars or different treatments. Explain why the randomisation of treatments is important. (1)

c) Apart from the addition of fertilisers, give **one** other method used in agriculture to increase crop yield. (1)

d) Explain why it is an advantage to grow crops on an area of fertile soil rather than to use the area as pasture for raising cattle. (1)

2 **a)** Crop breeders seek to develop crops with improved characteristics.
Give **two** examples of improved characteristics that would be useful in terms of increasing food security. (2)

b) Many crop plants are F_1 hybrids. Describe how these F_1 hybrids are produced and state the advantage gained by growers of these crops. (2)

3 In an investigation of the effects of different wavelengths of light on photosynthesis in green algal cells, apparatus was set up as shown in the diagram below. The glass tube containing a suspension of the cells was illuminated using filters to provide the different wavelengths of light.

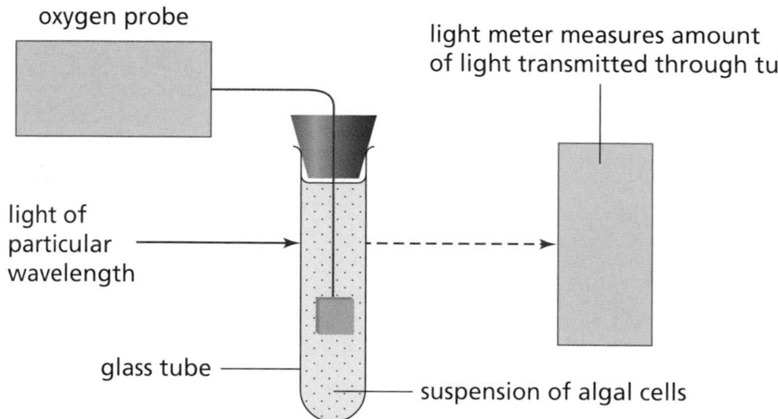

a) Identify **two** variables that should have been kept constant each time a different filter was used during the investigation. (2)

b) Describe a suitable control for this investigation. (1)

c) The light meter measures transmitted light. Apart from being transmitted, state **two** other possible effects of the algal cell suspension on light striking it. (2)

d) Describe the measurements that would need to be taken to determine the rate of photosynthesis in the algal cells at each wavelength tested. (2)

⇨

4 The bar chart below shows the effect of a selective breeding programme in cattle to increase milk yield over a period of 60 years. Error bars are also shown.

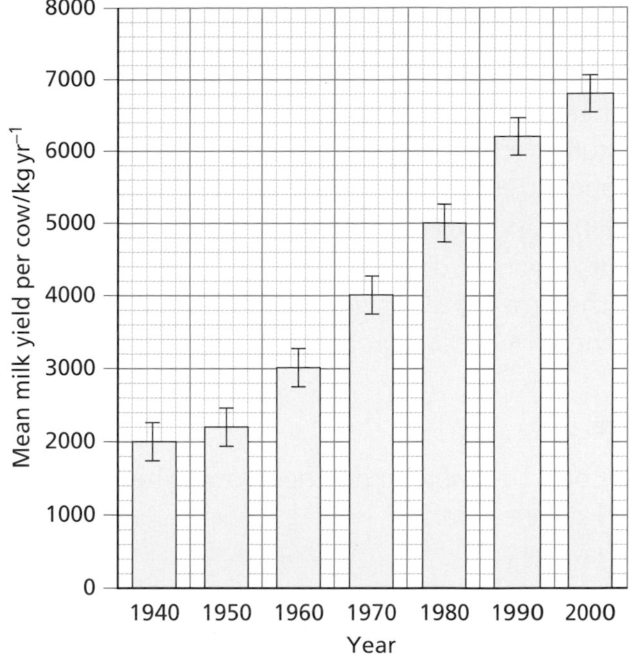

a) Describe what additional information the error bars provide about the milk yield per cow. (1)

b) Calculate the percentage increase in the mean milk yield per cow over the 60-year period. (1)

c) Other than the effect of selective breeding, suggest **one** reason that could account for the change in the mean milk yield over this period. (1)

d) In animals, individuals from different breeds might produce a new F_1 cross-breed population with improved characteristics. The improved characteristics could be lost in an F_2 generation due to the wide variety of genotypes that they would be expected to show.
Describe **one** process which would maintain the improved breed. (1)

e) A recent study has suggested that the increase in milk production has been accompanied by a reduction in the level of welfare concern for the cattle involved.
Give **two** examples of behavioural indicators of poor welfare. (2)

5 Insect pests of crop plants can be controlled by chemical pesticides, biological agents and by cultural practices.

a) Name the form of management that combines these different forms of control. (1)

b) Two-spotted mites are pests of strawberry plants. An investigation was carried out to provide information on the use of a predatory species of mite to control the two-spotted mites. Predatory mites were released on strawberry plants that were infested with two-spotted mites and the percentage of strawberry leaves occupied by both species was recorded over a 16-week period. The results are shown in the graph below.

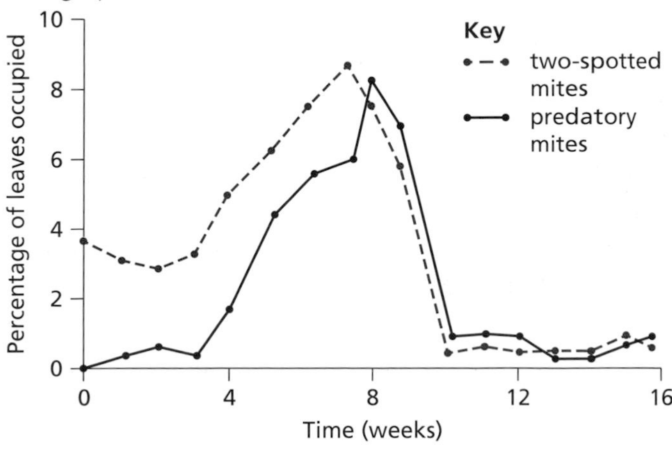

(i) Describe how the percentage of leaves occupied by predatory mites changed during the period of this investigation. (2)

(ii) The investigation was repeated but a systemic pesticide was applied to the strawberry plants after 10 weeks.
Predict the effect that the application of pesticide would have on both the two-spotted and predatory mite numbers. (1)

c) Give **one** example of a cultural control method that can be used to protect crops from pests. (1)

6 A genetic study of the European tree frog, *Hyla arborea*, in Denmark has indicated that severe habitat fragmentation has resulted in a loss in genetic diversity and inbreeding depression.

a) Describe the components of genetic diversity. (2)

b) Explain the meaning of the term *inbreeding depression*. (2)

c) State what is meant by the term *habitat corridor*. (1)

Question 7 contains a choice.

7 *Either* **A** Write notes on social behaviour under the following headings:

a) altruism and kin selection (4)

b) primate behaviour (5)

(total = 9)

or **B** Write notes on photosynthesis under the following headings:

a) photolysis (4)

b) the carbon fixation stage (5)

(total = 9)

Answers to Practice course assessment: Area 3

Paper 1

1 D, 2 B, 3 D, 4 A, 5 B, 6 A, 7 C, 8 D, 9 B, 10 C [1 each = 10]

Paper 2

1 a) 550 000 kg [1]

b) to eliminate bias when measuring the effects of different environments [1]

c) using high-yielding cultivars; protecting crops from pests/disease/competition/any named example, e.g. pesticide, biological control, others [any 1 = 1]

d) energy is lost between each trophic level [1]

2 a) higher yield; disease resistance; frost resistance; increased nutritional value (other answers possible) [any 2 = 2]

b) cross-breeding of different strains/cultivars/inbred lines; increase in yield *or* increase in vigour [1 each = 2]

3 a) volume of algal suspension; concentration of algal suspension; algal species; time for photosynthesis; distance of light source from algae; temperature [any 2 = 2]

b) repeat the experiment using a glass tube with no algae/distilled water [1]

c) reflected; absorbed [1 each = 2]

d) time period; volume of oxygen produced [1 each = 2]

⇨

4 a) the maximum and minimum values/range obtained for cattle in the sample; the variance of the data; values chosen to show range [any 1 = 1]
b) 240% [1]
c) better feeding; better animal welfare; genetic modification; other [any 1 = 1]
d) select parents from the original true-breeding stock *and* cross-breed them again [1]
e) stereotypic behaviour; failure of reproduction; failure of parental behaviour; altered activity levels; misdirected behaviour [any 2 = 2]

5 a) integrated pest management/IPM [1]
b) (i) increased up to eight weeks; decreased to 10 weeks *and* then remained constant [1 each = 2]
(ii) reduced number of/killed both species [1]
c) crop rotation; use of netting/scarecrows; (deep) ploughing; weeding [any 1 = 1]

6 a) the number; and frequency of all the alleles within a population [2]
b) an increase in the frequency of individuals who are homozygous; for recessive deleterious alleles [2]
c) an area of habitat connecting wildlife populations separated by human activities or structures [1]

7A a) donor harmed; recipient benefits; example of altruism; roles can be reversed during reciprocal altruism [any 2 = 2]
kin are close relatives who share genes; kin selection is donating resources to kin; kin selection increases the chances of the donor's own genes continuing in future generations [any 2 = 2]
[max = 4]
b) primates are monkeys, apes and humans; have long dependency period to learn complex social behaviour; ritualistic/appeasement behaviour designed to reduce conflict; examples of behaviour with description from grooming – mutual preening; facial expression/body posture/sexual presentation – acting as signals; alliances increase social status [any 5 = 5]
[total = 9]

B a) light energy excites electrons in pigments; splitting of water molecules in light-dependent stage; using energy from excited electrons; oxygen produced, which is released; hydrogen is accepted by NADP [any 4 = 4]
b) carbon dioxide fixed by RuBP; RuBisCo catalyses this reaction; 3-phosphoglycerate (3PG) converted to G3P; using ATP from light-dependent stage; using hydrogen from NADPH; sugars/glucose/carbohydrates made; RuBP regenerated [any 5 = 5]
[total = 9]

4 Skills of scientific inquiry

The questions within the Key Area chapters of this book test demonstration of knowledge. This section covers the Skills of scientific inquiry and includes questions to test these. We have given three different approaches to working with these science skills and recommend that you use all three.

4.1 gives an example of a scientific investigation and breaks it down into its component skills. There are questions on each skill area. The answers are on page 166.

4.2 goes through the skills one by one and gives you some exam-style questions covering the apparatus, techniques and skills with which you must be familiar. There is a grid on page 150 showing which skills are tested in the parts of each question. The answers are given on pages 167–168.

4.3 provides sets of hints and tips on answering science skills exam questions.

4.1 An example

Investigating the effects of different antibiotic concentrations on the growth of *E. coli*

Read through the information about the experiment below and then work through the skill areas listed, commenting on the questions in each category.

Introduction

Bacteria can grow into colonies on nutrient agar. Antibiotic multidiscs are blotting papers that have circular areas that have been soaked in antibiotic. Colonies are not able to grow in areas of agar into which antibiotic substances have diffused and so clear areas are created, as shown in the diagram below.

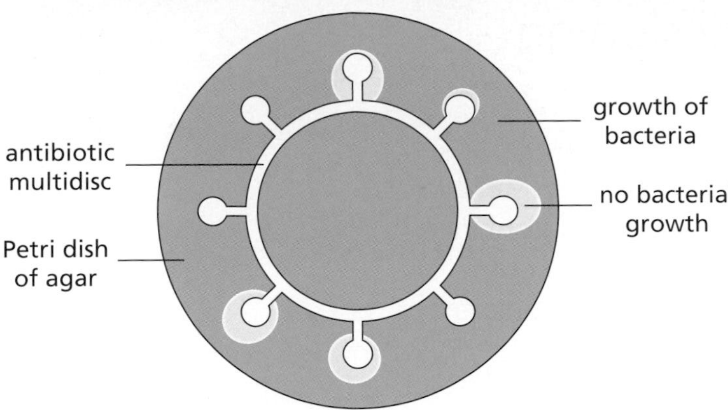

In an experiment that aimed to investigate the effects of increasing the concentration of an antibiotic on the growth of the bacterium *Escherichia coli* (*E. coli*), the following method was used. *E. coli* is an easily obtained, fast growing bacterium which is of great interest in medicine because it lives naturally in the human gut.

Method

1 Five Petri dishes were set up, each containing 10 ml of sterile nutrient agar.
2 An antibiotic multidisc with different concentrations of an antibiotic was added to the surface of the agar in each, pressed down and left for 24 hours.
3 Using a sterile pipette, 0.5 ml of broth containing a suspension of 10^5 *E. coli* per ml was removed from a culture bottle and squeezed onto the surface of each agar plate.
4 The culture was carefully spread across the agar surfaces using a glass spreader, which was sterilised between uses.
5 The dishes were kept in an oven at 30 °C for 48 hours, then removed, and the diameters of any clear zones that were visible around the antibiotic measured using a ruler.

Results

The mean diameters of the clear zones that developed around each antibiotic concentration are shown in the table below.

Antibiotic concentration (%)	Mean diameter (mm)
Control	0.0
0.5	0.1
1.0	3.3
1.5	4.1
2.0	4.6
2.5	6.1
3.0	6.6
3.5	6.7

Planning experiments and designing experiments

This is about confirming the aim of an experiment and suggesting a likely hypothesis, choosing apparatus, thinking about the dependent variable and deciding what to measure. Designing is closely related to planning but involves details of how often to measure, which variables need to be controlled and how to do this. It also involves anticipating possible errors and trouble-shooting these.

Q1 Suggest what the aim of the experiment was.

Q2 Suggest a hypothesis that would go with this aim.

Q3 Why is *E. coli* used?

Q4 What should be done to minimise the risks linked to using *E. coli* in the experiment?

Q5 What is the purpose of the control disc on the multidisc?

Q6 Why is nutrient agar used? (Hint: think about the conditions for bacterial growth.)

Q7 Why is the multidisc pressed into the agar and left for 24 hours?

Q8 Why is sterility important in this investigation? (Hint: many species of bacteria and fungi are present in air and on surfaces.)

Q9 Why is the plate left in the oven at 30 °C for 48 hours?

Q10 Why is the *E. coli* culture spread evenly on the agar?

Q11 Which variables have been controlled in this procedure?

Q12 Which is the dependent variable and which the independent variable?

Q13 Why was the experiment repeated five times?

Selecting information

This is about using a source such as a table or line graph to extract particular pieces of information. This can be simply reading off a value. The skill requires knowledge about labels, scales and units.

Q14 How could the results be expressed in a different numerical form?

Q15 Which concentration of antibiotic had the most effect on the *E. coli*?

Presenting information

This is about taking some information and presenting it in a different and more useful form. A common type of presentation might be to take information from a table and present it as a line graph. The skill requires knowledge of labels, scales and units, as well as careful drawing using a ruler and accurate plotting.

Q16 On a sheet of graph paper, draw a line graph to show the effect of antibiotic concentration on the growth of *E. coli*.

Processing information

This is often about working with numerical data and using calculations to convert a lot of data in a simple form.

Q17 What is the ratio of the diameter of the clear zone produced by antibiotic at 1% compared with that produced by antibiotic at 3%?

Q18 What is the mean diameter of the clear zones around the different antibiotic concentrations?

Q19 What is the percentage increase in the diameter of the clear zone at 0.5% antibiotic compared with 2.5%?

Q20 What is the range of diameters of clear zones in the investigation?

Q21 How many bacterial cells were in 0.5 ml of the culture liquid?

Q22 How could 0.5 ml of a culture with only 50% of this number be obtained?

Predicting and generalising

Predicting is about taking an experimental result and imagining what would happen if a variable changed. Generalising is about looking at experimental results and trying to find a rule that would hold true in all situations.

Q23 Predict the effect on the diameter of the clear zone produced if a concentration of the antibiotic at 4% was used.

Q24 How would you expect the results to compare if a different species of bacterium was used?

Q25 How could the diameters of the clear zones be increased without changing the concentration of antibiotic in the disc?

Q26 How could the diameters of the clear zones be decreased without changing the concentration of antibiotic in the disc?

Concluding and explaining

Concluding involves making a statement about the relationship between variables in an experiment. Explaining is about using knowledge to understand why a result has been obtained.

Q27 What is causing the clear zones?

Q28 What is the general conclusion that can be drawn from the results? (Hint: remember to look at the aim of the experiment.)

Evaluating

Evaluating is looking critically at an experiment and deciding if it is likely to be valid and if its result can be relied on; it is about looking for potential sources of error. Evaluating also involves suggesting improvements to an experiment that might remove sources of error in future experimental repeats.

Q29 What sources of error might be present?

Q30 How could the reliability of the results be improved?

Q31 How could this experiment be improved?

Q32 Which other substances that might affect bacterial growth could be investigated using a multidisc approach? How could you adapt the experimental set-up to investigate one of these?

Answers are on page 166.

4.2 Skill-by-skill questions

The table below lists the basic skill areas that can be tested in your exam and the techniques with which you must be familiar. We have provided six practice questions (Q1–6) which cover all of the skills areas and all of the techniques that you must know about for your exam. The table below shows the parts of the practice questions where you can find testing of each skill. It is probably better to try the whole of each question in turn. If you find particular difficulty with any part of a question, you could use the table to identify the skill area that needs further work.

Skill area	Category within skill area	Techniques questions					
		Q1	Q2	Q3	Q4	Q5	Q6
		Effect of inhibitor concentration	Using a respirometer	Carrying out chromatography	Gel electrophoresis	Measuring metabolic rate	Using a spectroscope
1 Selecting information …	from a line graph					a)(i)	a)(i), a)(ii), c)
	from a diagram				a)(i), a)(iii)		
2 Presenting information …	as a line graph	c)	b)				
3 Processing information …	as a ratio				a)(ii)		
	as an average			b)(i)			
	as a percentage					a)(ii)	a)(iii)
	by general calculation [+ − × ÷]		d)		b)(ii)	b)(i), b)(ii), c)	
4 Planning and designing	Planning: variables and control	a), e)	a)(ii)	a)(i), a)(iv)			
	Designing: apparatus and procedure	b)	a)(i), e)	a)(ii), a)(iii)		d)	a)(iv)
5 Predicting and generalising	Predicting				b)(iii)		d)(ii)
	Generalising					b)(iii)	
6 Concluding and explaining	Concluding	d)	c)	b)(ii)	b)(i)	a)(iii)	d)(i)
	Explaining						b)
7 Evaluating	Identifying source of error			b)(iii)1)			
	Suggesting improvement	f)		b)(iii)2)			
Total marks per question		**9**	**8**	**9**	**7**	**9**	**11**

Experimental questions

These questions use a selection of techniques that you need to know about for your exam. Answers are on page 167.

Question 1 Effect of inhibitor concentration on enzyme activity

Catechol oxidase is an enzyme that is found in banana tissue. It is involved in a two-step reaction which produces the brown melanin pigment that forms in cut or damaged fruit.

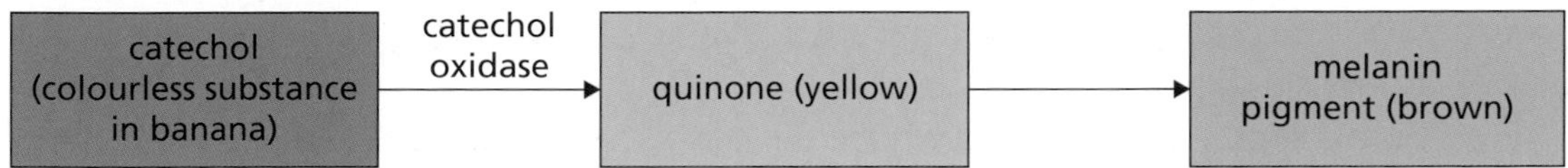

The effect of the concentration of lead ethanoate on this reaction was investigated.

20 g of banana tissue was cut up, added to 20 cm^3 of distilled water and then liquidised, filtered and a buffer added to keep the pH constant during the reaction. This produced a buffered extract containing both catechol and catechol oxidase.

Test tubes were set up as described in **Table 1** and kept at 20 °C in a water bath.

Table 1

Test tube	Contents of test tubes
A	sample of extract + 1 cm^3 0.01% lead ethanoate solution
B	sample of extract + 1 cm^3 0.1% lead ethanoate solution

Every 10 minutes, the tubes were placed in a colorimeter which measured how much brown pigment was present. The more brown pigment present, the higher the colorimeter reading.

The results are shown in **Table 2**.

Table 2

Time (minutes)	Colorimeter reading (units)	
	Test tube A	Test tube B
	sample of extract + 0.01% lead ethanoate	sample of extract + 0.1% lead ethanoate
0	1.8	1.6
10	5.0	2.0
20	6.0	2.2
30	6.4	2.4
40	7.0	2.4
50	7.6	2.4
60	7.6	2.4

a) **(i)** Identify the independent variable in this experiment. [1]

(ii) Identify **two** variables, not already mentioned, that would have to be kept constant. [2]

b) Explain why the initial colorimeter readings at 0 minutes were not 0.0 units. [1]

c) On a separate piece of graph paper, copy and complete the graph below to show all of the results of the experiment. [2]

⇨

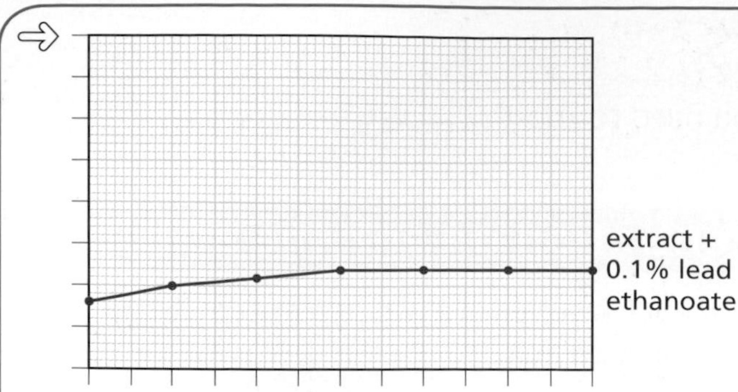

d) Describe the effect of the concentration of lead ethanoate solution on the activity of catechol oxidase. [1]
e) Describe a suitable control for this investigation. [1]
f) State how the procedure could be improved to increase the reliability of the results. [1]

Question 2 Using a respirometer

An experiment was carried out to compare the rate of metabolism in a species of woodlouse at different temperatures.

Five woodlice were placed in a respirometer flask which was fitted with a carbon dioxide (CO_2) probe linked to a computer as shown in the diagram below.

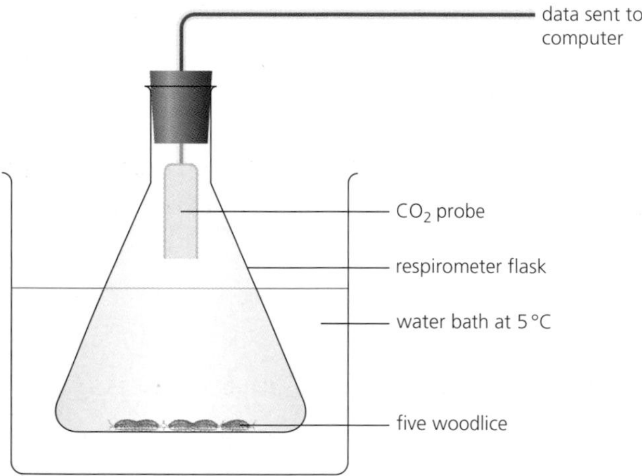

The flask was placed in a water bath at 5 °C and left for 10 minutes.

The CO_2 produced per minute was then measured. This procedure was repeated at different temperatures and the results shown in the table below.

Table

Temperature (°C)	Rate of CO_2 production (units per minute)
5	150
10	250
20	600
25	700
30	800

a) (i) Give a reason why the flask was left for 10 minutes at each temperature **before** each reading was taken. [1]
(ii) A control flask should be included in this investigation.
Describe the control which should be included and explain its purpose in the investigation. [2]
b) On a separate piece of graph paper, plot a line graph to show the results of this experiment. [2]
c) Draw a conclusion from these results. [1]
d) Calculate the volume of CO_2 produced by one woodlouse kept at 20 °C for 2 minutes. [1]
e) Identify a feature of the method which improved the reliability of the results obtained. [1]

Question 3 Carrying out chromatography

In an investigation into photosynthetic pigments found in two different tomato cultivars, a procedure was carried out as described below.

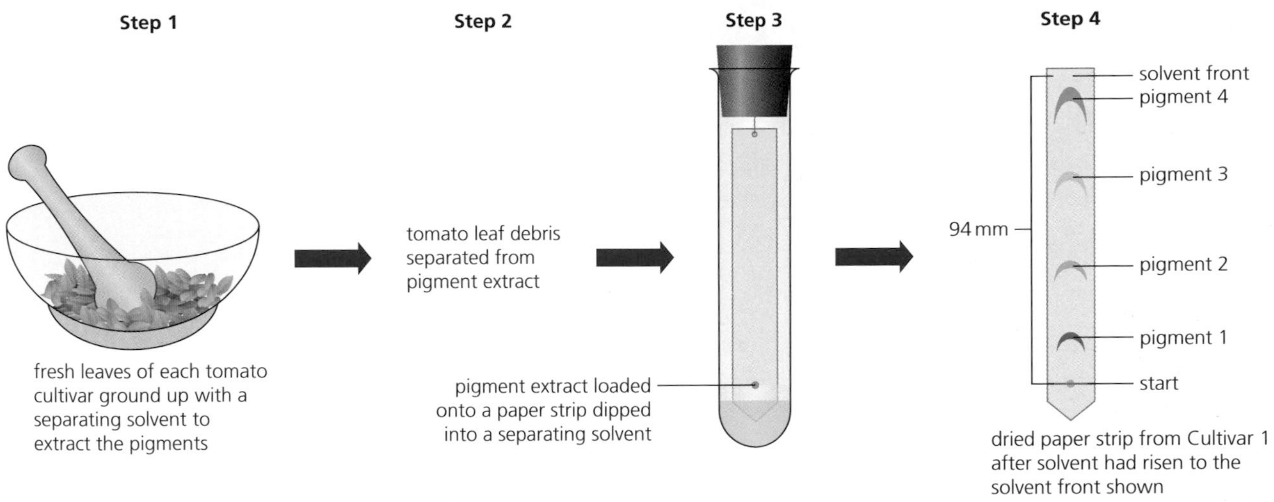

a) (i) Identify **two** variables related to **Step 1** of the procedure which should be kept the same for each tomato cultivar. [2]

(ii) Describe a method which could be used to separate debris in **Step 2**. [1]

(iii) Name the technique used to separate the pigments in **Step 3**. [1]

(iv) Give one variable related to **Step 3** which should be kept the same for each tomato cultivar. [1]

b) Pigments can be identified by their relative front (R_f) values which are calculated by the following formula.

$$R_f = \frac{\text{distance travelled by pigment front}}{\text{distance travelled by solvent front}}$$

The table below shows the R_f values for the four pigments which were extracted from each tomato cultivar.

	R_f value	
Pigments	**Cultivar 1**	**Cultivar 2**
1	0.25	0.25
2	0.52	0.51
3	0.68	0.67
4	0.92	0.93

(i) Calculate the distance travelled by pigment 3 front from cultivar 1. [1]

(ii) Suggest a conclusion which could be drawn about the pigments found in cultivars 1 and 2. [1]

(iii) Pigment 2 was identified as chlorophyll a.

1 Suggest a reason why the R_f values obtained for chlorophyll a from the two cultivars are different. [1]

2 Suggest an improvement to the method which might result in closer values for the same pigments being obtained. [1]

Data questions

These questions refer to a selection of the techniques that you need to know about for your exam. Answers are on page 167.

Question 4 Carrying out gel electrophoresis

In an investigation into the evolution of frogs, samples of proteins were extracted from four frog species. The samples were stained and added to wells in a block of agarose gel. An electric current was passed through the gel block and the different proteins in each sample passed along the block to different extents. The final position of the proteins are shown by the stained bands in the gel.

The results are shown in **Figure 1**.

Figure 1

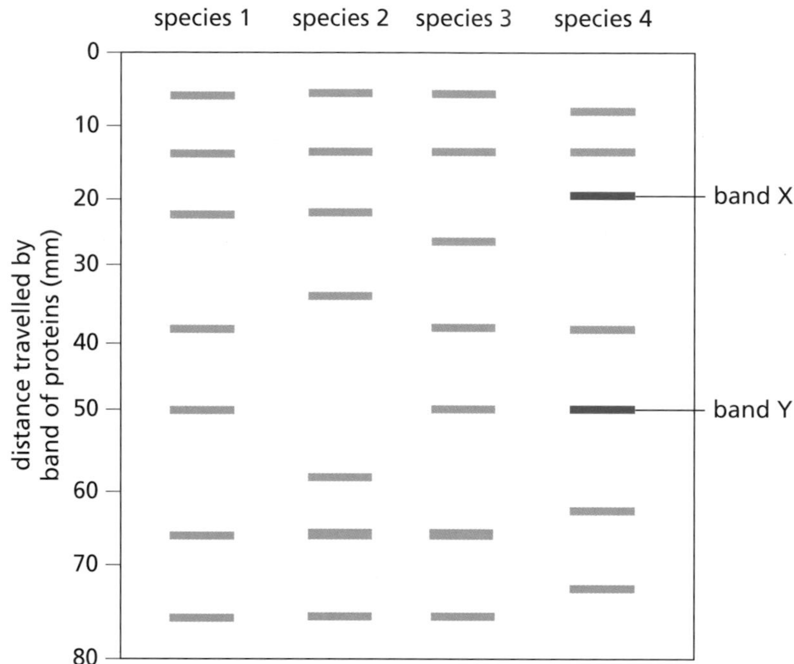

The results of the investigation were used, along with other evidence, to construct a proposed phylogenetic tree of the four frog species as shown in **Figure 2**.

Figure 2

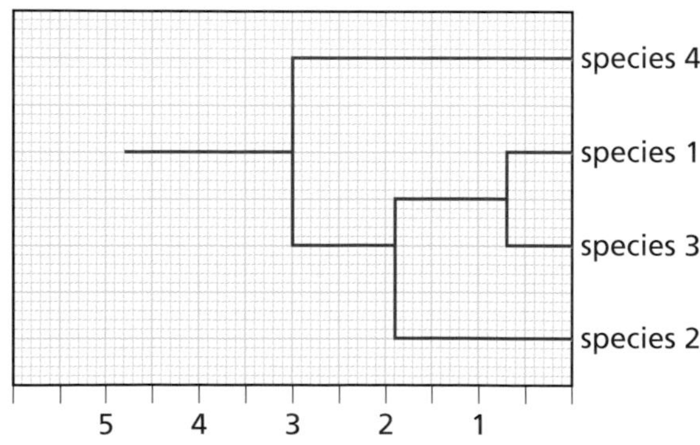

a) (i) Describe the general similarities and difference between the proteins from species 1 compared with species 2. [2]

(ii) Calculate the simplest whole number ratio of the distance travelled by band X to that travelled by band Y. [1]

(iii) Give the number of proteins which species 2 has in common with species 3. [1]

b) (i) Give evidence from Figure 1 which supports the proposed position of species 4 in the phylogenetic tree. [1]

(ii) From the phylogenetic tree, calculate how long ago the last common ancestor of species 4 and species 3 existed. [1]

(iii) A fossil bone was discovered which was proposed to be from the last common ancestor of species 1 and 2.

If the proposal was true, predict the age of the fossil bone. [1]

Question 5 Measuring metabolic rate

Ruby-throated hummingbirds *Archilochus colubris* have very high metabolic rates during normal activity. They migrate thousands of kilometres each year between their wintering areas in Central America and their summer breeding areas in parts of North America.

They feed on nectar throughout the year and save energy in cold conditions during the night by entering a temporary state known as torpor, during which their metabolic rates are much reduced.

Graph 1 shows how the average body masses of the birds changed over a yearly cycle and Graph 2 shows how the birds' average metabolic rate during normal activity and torpor changed with air temperature.

Graph 1

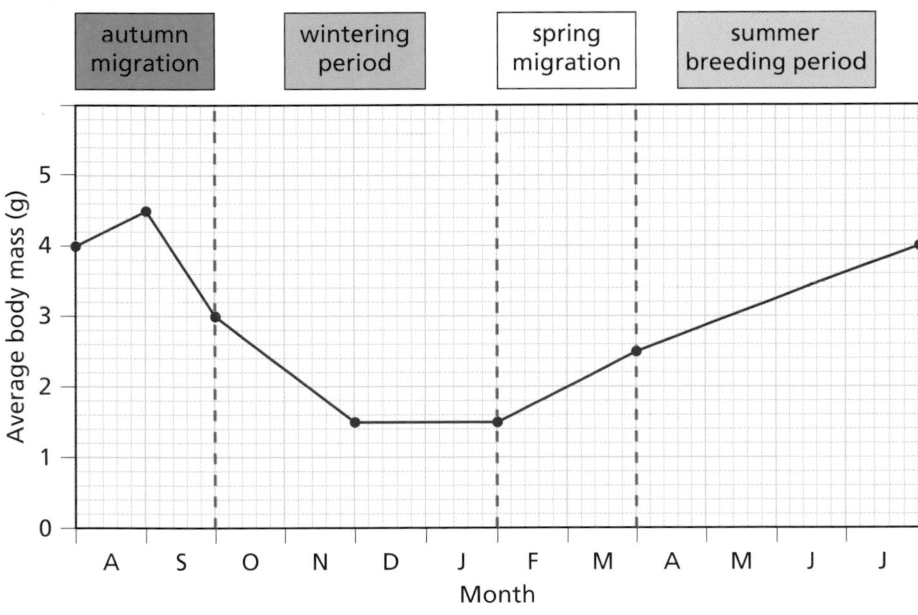

Graph 2

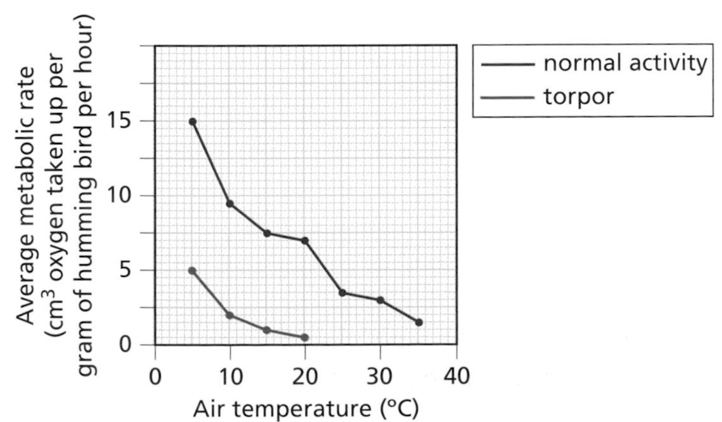

a) (i) **Use values from Graph 1** to describe the changes in average body mass of the hummingbirds from the beginning of August until the end of January. [2]

(ii) Calculate the percentage increase in average body mass during the summer breeding period. [1]

(iii) Suggest **one** reason for the increase in body mass of the birds during summer. [1]

b) (i) Calculate the average decrease in oxygen consumption per gram per hour for each degree reduction in air temperature from 5 °C to 35 °C during normal activity. [1]

(ii) Calculate the reduction in metabolic rate of a bird which enters torpor from normal activity at 10 °C. [1]

(iii) Suggest why no values are given for metabolic rate during torpor above 20 °C. [1]

c) Using information from the **Graphs 1** and **2**, calculate the volume of oxygen consumed per hour by a hummingbird, at the end of September, during normal activity at 20 °C. [1]

d) Apart from using oxygen consumption per hour, give **one** other measure of metabolic rate. [1]

Question 6 Using a spectroscope

As part of an investigation into the productivity of a wheat variety, 10 m^2 plots in a large greenhouse were planted with equal numbers of wheat plants and kept constantly illuminated with light of different wavelengths. The rate of photosynthesis was measured in each plot and the results are shown in **Graph 1**.

Chlorophyll a was extracted from a sample of wheat leaves. The absorption spectrum of the chlorophyll a extract was measured and is shown in **Graph 2**.

White light was shone through the chlorophyll a extract then passed into the slit of a hand-held spectroscope. The resulting spectrum is shown in **Figure 1.**

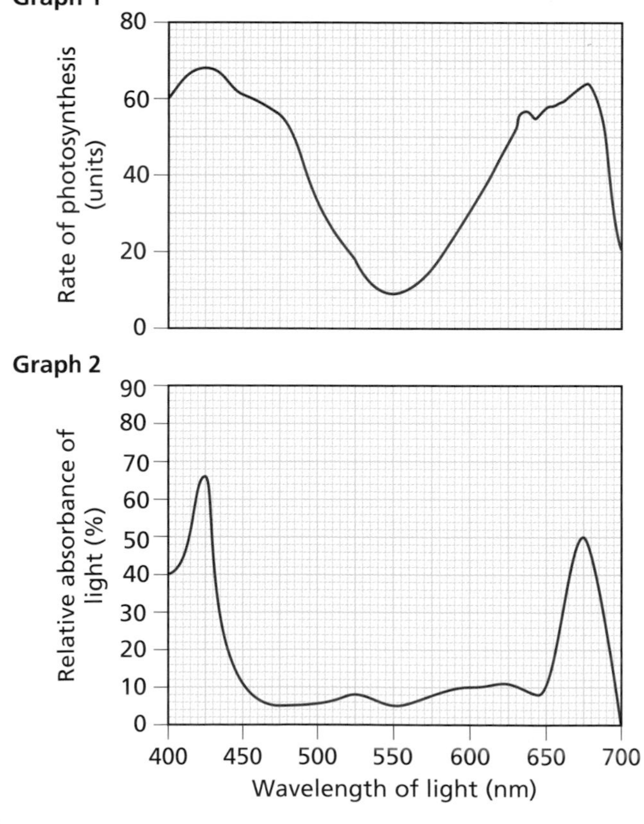

Figure 1

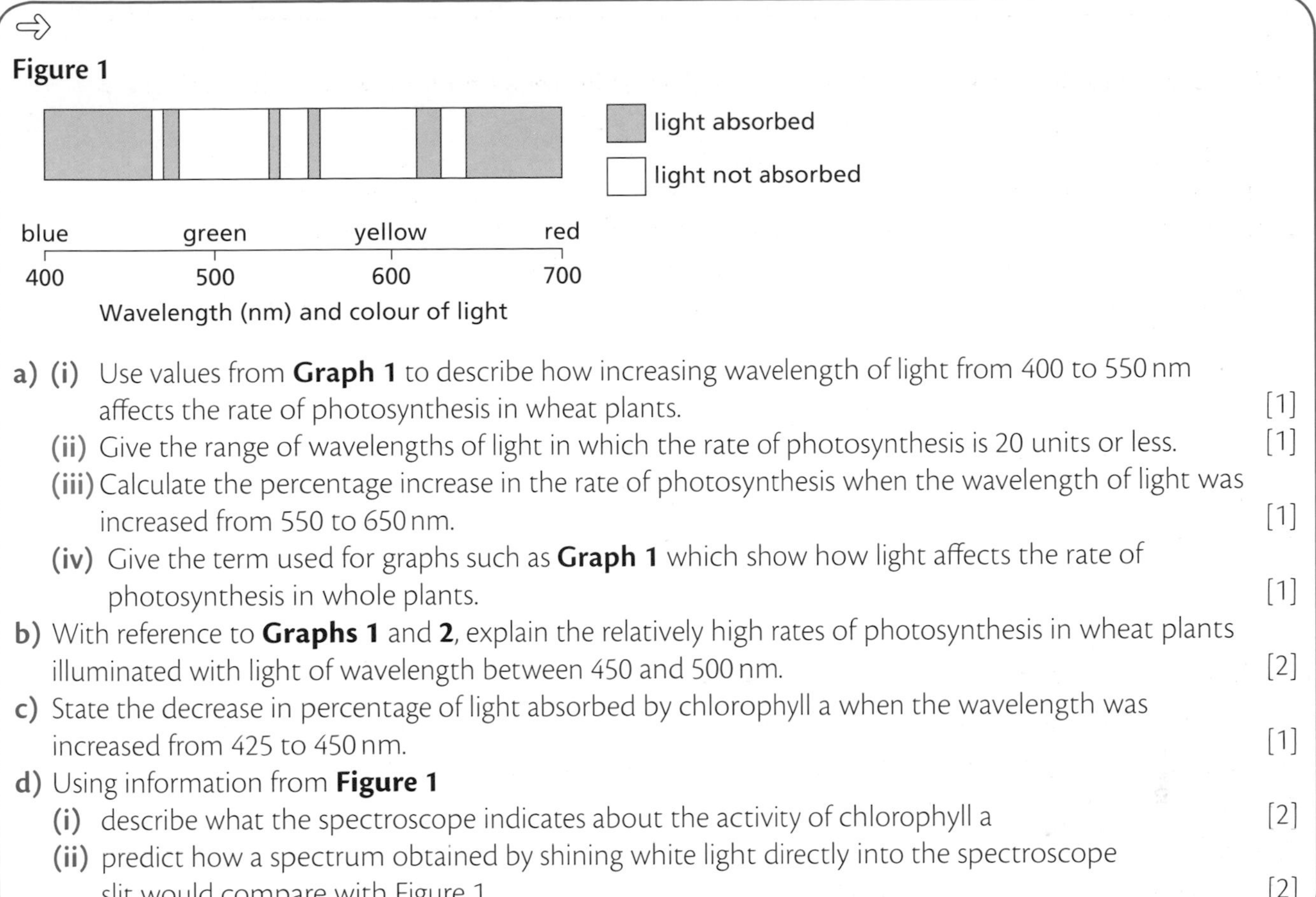

a) (i) Use values from **Graph 1** to describe how increasing wavelength of light from 400 to 550 nm affects the rate of photosynthesis in wheat plants. [1]

(ii) Give the range of wavelengths of light in which the rate of photosynthesis is 20 units or less. [1]

(iii) Calculate the percentage increase in the rate of photosynthesis when the wavelength of light was increased from 550 to 650 nm. [1]

(iv) Give the term used for graphs such as **Graph 1** which show how light affects the rate of photosynthesis in whole plants. [1]

b) With reference to **Graphs 1** and **2**, explain the relatively high rates of photosynthesis in wheat plants illuminated with light of wavelength between 450 and 500 nm. [2]

c) State the decrease in percentage of light absorbed by chlorophyll a when the wavelength was increased from 425 to 450 nm. [1]

d) Using information from **Figure 1**

(i) describe what the spectroscope indicates about the activity of chlorophyll a [2]

(ii) predict how a spectrum obtained by shining white light directly into the spectroscope slit would compare with Figure 1. [2]

4.3

Hints and tips

Tips on selecting information

Note that you can be asked to deal with information that is more complex than that which you could be asked to present or process.

- Some questions might be based on passages of *text* – you could be asked to pick out information, identify evidence, explain relationships, draw conclusions and show biological knowledge. Try using a highlighter to pick out important points in the text.
- Data might be presented as a *table* – again, a highlighter is helpful.
- Figures in tables may have + or – symbols to indicate if observed values are significantly higher or lower than would have been expected by chance.
- The main types of chart you could see are **bar charts** and **pie charts**.
- You could be presented with **line graphs** or **graphs of best fit**. A line graph is used when data points are simply connected by straight lines. If data are scattered on a graph, lines can be fitted using the results of calculations on the data.
- On graphs, the variable being investigated (independent variable) is usually on the *x*-axis and what is being measured (dependent variable) is on the *y*-axis.
- You might be asked to identify variables that should be controlled – use **CID**:
  - **C**ontrolled variables should be kept **C**onstant.
  - **I**ndependent variables are being **I**nvestigated.
  - **D**ependent variables give the **D**ata that form the results.
- Sampled data, either from fieldwork or from time intervals, are often graphed on the *x*-axis but are not true variables. Watch out for this if you are asked to identify a variable.
- Watch out for graphs with a double *y*-axis – these are tricky! The two *y*-axes often have different scales to increase the difficulty. You must take care to read the question and then the graph carefully to ensure that you are reading the correct *y*-axis – there will usually be a key, which is critical.
- On bar charts and line graphs, work out the value of the smallest square on either scale before trying to read actual values.
- You might see a graph with a **semi-logarithmic scale**. These are often used when the numbers involved in a graph scale range from low to very high, such as those dealing with numbers of bacteria present in a growing culture. An example is given on page 82.
- You could see bar charts or line graphs with **error bars**. These are used to show the extent of variability of data, the level of confidence that exists regarding the data or if two sets of data differ from each other significantly. You would not be asked to draw error bars.

- If you are asked to calculate an increase or decrease between points on a graph, you should use a ruler to help accuracy – draw pencil lines on the actual graph if this helps.
- When you are asked to describe a trend, it is essential that you quote the values of the appropriate points and use the exact labels given on the axes in your answer. You must use the correct units in your description.
- Sometimes there will be two sources of data, for example a graph and a table. Make sure you study the two sources carefully – there will be something that links them and it is this link that you will be asked to use.
- You could be asked to deal with statistical measures such as **mean**, **range** and **standard deviation**. The mean is an arithmetic average of the data and the range is the difference between the highest and lowest values in a group. The standard deviation is a measure of how varied the data are. You would not be expected to calculate standard deviation values.
- **Box plots** are used to show differences between groups of data – they are graphical ways to display the data so that groups can be compared visually. A group of data is put into rank order. The rank is split into four quartiles, each of which contains 25 per cent of the items of data. The value at the boundary of the second and third quartile is called the median value. The box plot for a group of unspecified data values is constructed as shown below.

Ranked data values	*y*-axis scale	Box plot	Quartile
204	200		
206	202		
207	204		lowest value in data
210	206		first quartile in data
212	208		
213	210		second quartile in data
214	212		median
217	214		
218	216		third quartile in data
219	218		fourth quartile in data
220	220		highest value in data

- You could see box plots in a question but you would not be asked to draw one.
- **Keys** are used to identify and, usually, name an organism using features of its body as clues.

Tips on presenting information

- The most common question in this area requires students to present information that has been provided in a table as a graph – usually a line graph or sometimes a chart (usually a bar chart), although it is possible that you could be asked for a simple diagram or to draw a key.
- Check the question to see if it is a line graph or a bar chart that is required – the question will usually tell you.
- Marks are given for providing scales, labelling the axes correctly and plotting the data points. Line graphs require points to be joined with straight lines using a ruler. Bar charts need to have the bars drawn precisely, using a ruler.
- Ensure that you can identify the dependent and independent variables.
- The graph labels should be identical to the table headings and units. Copy them exactly, leaving nothing out.
- You need to decide which variable is to be plotted on each axis. The data for the variable under investigation (independent variable) is usually placed in a left column of a data table and should be scaled on the *x*-axis. The right column in a data table usually provides the label and data for the *y*-axis – the dependent variable. You will lose a mark if these are reversed.
- You must select suitable scales. Choose scales that use at least half of the graph grid provided, otherwise a mark will be deducted. The values of the divisions on the scales you choose should allow you to plot all points accurately.
- Make sure that your scales include zero if appropriate and extend beyond the highest data points.
- The scales must rise in regular steps. At Higher level the examiners will often test you on this by deliberately skipping one of the values that they have given you to plot in the table.
- Be careful to include one or both zeros on the origin if appropriate. It is acceptable for a scale to start with another value other than zero if this suits the data.
- Take great care to plot the points accurately using crosses or dots and then connect them exactly using a ruler.
- Don't plot zero or connect the points back to the origin unless zero is actually included in the data table. If zero is there, you must plot it.
- When drawing a bar graph, ensure that the bars are the same width. Remember to include a key if the data require it.
- If you make a mistake in a graph, a spare piece of graph paper is provided at the end of your exam paper.

Tips on processing information

General points

- You can be asked to do calculations involving whole numbers, decimals or fractions. Answers might be whole numbers or decimals.
- Decimal answers should be rounded to the appropriate degree of accuracy, which will generally be to the nearest two decimal places or to three significant figures.
- You can be asked to convert between units, such as those for mass (μg, mg, g and kg) or for distance (μm, mm, m and km).
- You might be asked to do calculations involving negative numbers or using scientific notation.
- You might be asked to put values into a given equation and calculate an unknown.

Tackling the common calculations

Percentages: expressing a number as a %

The number required as a percentage is divided by the total and then multiplied by 100, as shown:

$$\frac{\text{number wanted as a \%}}{\text{total}} \times 100$$

Percentage change: increase or decrease

First, calculate the increase or decrease to find the change. Then, express this value as a percentage:

$$\frac{\text{change}}{\text{original starting value}} \times 100$$

Ratios

These questions usually require you to express the values given or being compared as a simple whole number ratio.

First you need to obtain the values for the ratio from the data provided in the table or graph. Take care that you present the ratio values in the order they are stated in the question. Then simplify them, first by dividing the larger number by the smaller one then dividing the smaller one by itself. However, if this does not give a whole number then you need to find another number that will divide into both of them. For example, 21:14 cannot be simplified by dividing 21 by 14 since this would not give a whole number. You must then look for another number to divide into both, in this case 7. This would simplify the ratio to 3:2, which cannot be simplified any further.

Mean

This is one type of average – the others are the median (used in box plots and indicating the middle value of a set of data) and the mode (the most common value in a set of data).

Add up the values provided and then divide the total by the number of values given. Make sure you include all values even if one or more is actually a zero value – they still count.

You might be asked to calculate the mean increase per unit time in a value over a period. If so, calculate the total increase, then divide by the number of units of time in the period given.

Range

This refers to the difference between the lowest and the highest values in a set of data. Find the lowest and the highest – subtract one from the other.

Tips on experimental skills of planning, designing and evaluating

Experimental aims

The aim of an experiment or investigation is usually stated at the start of a question – make sure you read the stem carefully because you cannot give a conclusion without knowing the aim.

Hypothesis formation

When an observation is made, the suggested scientific explanation for it is called a hypothesis. You may be given an observation and asked to construct a hypothesis.

Variables

You should be able to identify the independent and dependent variables in an investigation or experiment.

- The **i**ndependent variable is the **i**nvestigated variable – it usually appears in the first column of a data table and is plotted on the *x*-axis of a graph.
- The **d**ependent variable refers to the **d**ata (results) produced – it usually appears in the second column of a data table and is plotted on the *y*-axis of a graph.

Validity

To achieve validity, only the independent variable should be altered while the other variables should be kept constant. Examples of the variables that might need to be controlled and kept constant to ensure results are valid include temperature, pH, concentration, mass, volume, length, number, surface area and type of tissue, depending on the experiment.

The control should be identical to the original experiment apart from the one factor being investigated. If you are asked to describe a suitable control, make sure that you describe it in full. A control experiment allows a comparison to be made and allows you to relate the dependent variable to the independent one.

Negative controls are set up to be sure that the variable under investigation is causing the result. If a new fertiliser were being tested in a field trial, plots without the fertiliser would act as negative control. Any difference between the fertiliser plots and control plots would be due to the fertiliser. Positive controls are set up to check the experimental method. If a new fertiliser were being tested in a field trial, plots with an existing fertiliser would act as positive controls. If the existing fertiliser plots did not have enhanced crop growth, there must be something going wrong with the cultivation methods being used.

Experimental procedure

- If the effect of temperature on enzyme activity is being investigated, it is good practice to allow solutions of enzyme and substrate to reach the required temperature before mixing them, to ensure that the reaction starts at the experimental temperature.
- It is good experimental practice to use percentage change when you are comparing results. A percentage change allows a fair comparison to be made when the starting values in an investigation are different.
- Watch out for questions that refer to dry mass. Since the water content of tissues is variable and can change from day to day, dry mass is often used when comparing masses of tissues that are expected to change under experimental conditions.
- Precautions to minimise errors include washing apparatus such as beakers or syringes or using different ones if the experiment involves different chemicals or different concentrations. This prevents cross-contamination.
- Questions regarding procedures that ask why the experiment was left for a certain time require you to state that this is to allow enough time for particular events to occur. These events could include the following:
  - diffusion or absorption of substances into tissue
  - growth taking place
  - the effect of substances being visible
  - a reaction occurring.

Observations and measurements

When observing and measuring, you need to ensure reliable results. To improve the reliability of experiments and the results obtained, the experiment should be repeated.

Remember **ROAR**: **r**epeat and **o**btain an **a**verage, which increases **r**eliability.

Modifications needed in light of experience

You will probably be asked to suggest a modification to an experimental procedure to test different variables. If you are asked about this, think about how to alter the different variables while keeping the original variable constant. For example, when investigating enzyme action, temperature is often varied. If temperature were kept constant, then pH level could be investigated as long as all other variables were kept the same.

Concluding, predicting and generalising

When **concluding**, you must provide a reference to the experimental aim, which is likely to be stated in the stem of the question. You could be asked to:

- summarise experimental results, including describing patterns, trends or rates of change
- look at supplied information and **predict** results or outcomes of experiments
- make **generalisations** about, state relationships between, or suggest rules about biological processes.

Key words

Accuracy – the extent to which a measurement is near to its true value
Balance – instrument for measuring the weight or mass of materials
Beaker – container to hold and pour liquids for use in experiments
Colorimeter – device used to measure the concentration of a coloured solution
Control – technique which can help to ensure that experimental results are caused by the independent variable
Dependent variable – variable which forms the results of an experiment
Dropping pipette/syringe – instrument for delivering small measured volumes of liquid
Funnel – device for channelling liquids or gas bubbles in an experiment; can be used with a filter paper to separate mixtures of solid and liquid
Independent variable – variable being investigated and so purposely varied
Measuring cylinder – container to hold and measure volumes of liquid
Reliability – measure of confidence in the results of an experiment
Source of error – a factor or feature of experimental procedure which can lead to mistakes
Spectroscope – device which shows the spectrum of light coming from a source
Stopwatch – instrument for measuring time in an experiment
Test tube – glass container for holding liquid mixtures in experiments
Thermometer – instrument for measuring temperature in an experimental situation
Validity – measure of the fairness of an experiment
Variable – factor in an experiment which can change
Water bath – trough of thermostatically heated water to provide steady, controlled temperatures

Scientific inquiry skills answers

4.1 An example

Q1 To show the effects of increasing antibiotic concentration on the growth of *E. coli*

Q2 Increasing the concentration of antibiotic would increase the inhibition of growth of *E. coli* *or* converse *or* increasing the concentration of antibiotic would have no effect [any 1]

Q3 Easily obtained bacterium which grows rapidly in culture

Q4 A risk assessment should be carried out to identify and minimise the impact of hazards using sterile/aseptic techniques

Q5 To show that any effect recorded is due to presence of antibiotic

Q6 Provides food and other requirements for the culture of *E. coli*

Q7 To allow the substance to diffuse into the agar

Q8 To prevent entry of other microorganisms that might compete with *E. coli* and affect results *or* that are hazardous

Q9 To provide optimum conditions for enzymes involved in growth of *E. coli*

Q10 So that any colonies develop evenly on the agar and are easier to see

Q11 Volume of nutrient agar; volume of *E. coli* culture; concentration of *E. coli* culture; temperature; time dishes were left for; type of antibiotic; area of disc in contact with agar

Q12 independent – concentration of antibiotic; dependent – diameter of clear zones

Q13 To increase the reliability of the results

Q14 Area/radius of clear zones

Q15 3.5%

Q16 Scales, labels and units *and* points plotted accurately and joined with straight lines

Q17 1:2

Q18 4.5

Q19 6000%

Q20 6.6 mm

Q21 50 000 *or* 5×10^4

Q22 Mix the culture with an equal volume of distilled water

Q23 6.7–6.8 mm

Q24 Might be expected to have similar effects

Q25 Leave dishes longer at start to allow more antibiotic to diffuse into agar

Q26 Do not leave dishes for 24 hours at the start of the experiment

Q27 Antibiotic prevents growth of bacteria, so no colonies develop

Q28 That increasing the concentration of antibiotic increases inhibition of the growth of *E. coli*

Q29 Potential for contamination; failure to have multidisc evenly pressed down; failure to spread bacteria evenly; errors in measuring diameters of clear areas

Q30 Reliability could be increased by repeating with more dishes

Q31 More dishes; more accurate measurement of clear zones

Q32 Different types of antibiotic; nutrients; precursors; inducers; hormones; pesticides; others; repeat experiment with discs containing appropriate substances but keeping all other variables constant

Scientific inquiry skills answers

4.2 Skill-by-skill answers

Question			Answer
1	a	i	Concentration of lead ethanoate
		ii	Volume of extract; volume/mass of buffer/variety of banana (any 2 = 2)
	b		The reaction had already started
	c		Correct scales and labels from data table headings = 1 Accurate plotting and straight-line connection of points and labelling of line = 1
	d		The greater the concentration of lead ethanoate the greater the inhibition of the enzyme
	e		A tube with a sample of extract and 1 cm^3 distilled water
	f		Repeat the entire procedure at each of the lead ethanoate concentrations (and average the results)
2	a	i	To allow the flask to come to/respiration to occur at the appropriate temperature
		ii	A flask at each temperature without woodlice/with glass beads instead of woodlice/with dead woodlice = 1 To ensure that it was respiring woodlice which caused the results = 1
	b		Correct scales and labels from data table headings = 1 Accurate plotting and straight-line connection of points = 1
	c		The higher the temperature, the higher the metabolic rate of the woodlice
	d		240 units
	e		Using five woodlice (at each temperature)
3	a	i	Mass of leaves used/volume of solvent used/time of grinding (any 2)
		ii	Pass the ground-up material through a filter paper/muslin cloth
		iii	Paper chromatography
		iv	Temperature/volume of solvent/mass of extract loaded onto paper
	b	i	63.4 mm
		ii	That the cultivar contains the same/same number of pigments
		iii	Human error in measuring the solvent fronts Use thin layer chromatography
4	a	i	Species 1 has 2 proteins not found in species 2 = 1 Species 2 has two pigments not found in species 1 = 1
		ii	2:5
		iii	4
	b	i	Species 4 has the more protein differences compared with the other three species than the other species do compared with each other
		ii	300 million years before present
		iii	190 million years before present
5	a	i	Increased from 4.0 to 4.5 g from beginning to end of August = 1 Decreased from 4.5 to 1.5 g from end August to end November = 1
		ii	60%
		iii	Much food available OR less energy used in flying/migrating
	b	i	0.45 cm^3 oxygen per gram per hour
		ii	7.5 cm^3 oxygen uptake per g per hr
		iii	Birds do not enter torpor at high temperatures
	c		21 cm^3
	d		CO_2 production per hour

6	a	i	From 400–425 nm photosynthesis increases from 60 to 70 units = 1 From 425–550 nm photosynthesis decreases from 70 to 10 units =1
		ii	From 525 to 575 nm OR 50 nm
		iii	500%
		iv	Action spectrum
	b		Other pigments than chlorophyll a are absorbing light energy
	c		55%
	d	i	Absorbs highest in the blue and red areas of the spectrum = 1 and lowest in green (and yellow) areas = 1
		ii	All wavelength/colours of light would be visible/not absorbed

Your Assignment

The Assignment is based on an investigative research task that you have carried out in class time. You choose the topic to be studied and then investigate and research it. The Assignment must be an individually produced piece of work, although experimental and fieldwork can be done in small groups and the data shared.

The Assignment assesses the application of Skills of scientific inquiry and some related biology knowledge and understanding that you develop as you work through the Course.

You will have to write up the work in the form of a **Report** under controlled assessment conditions. During your write up, you will have access to your notes but **not** to a draft copy of any part of your final Assignment Report.

The Report will be marked out of 20 marks with 16 of the marks being for scientific inquiry skills and 4 marks for the demonstration and application of knowledge of background biology.

The Report is marked by SQA and scaled to 30 marks, contributing 20% to the overall grade for your Course.

The Assignment is supervised by your teachers, who should supply you with the *Instructions for Candidates* document published by SQA. Your Assignment has **two** main stages which can be split up into tasks, as shown in the table below.

<table>
<tr><th colspan="2">Stage</th><th colspan="2">Task</th><th>Recommended timing</th><th>Supervision level</th></tr>
<tr><td rowspan="3">1</td><td rowspan="3">Research</td><td>A</td><td>Selecting a topic; devising an aim; planning experiments or fieldwork; summarising the background biology</td><td rowspan="3">6 hours</td><td rowspan="3">Some supervision and control</td></tr>
<tr><td>B</td><td>Carrying out experiments or fieldwork; collecting numerical results; drawing conclusions; evaluating your work</td></tr>
<tr><td>C</td><td>Selecting information from existing sources to compare to experimental or fieldwork results</td></tr>
<tr><td>2</td><td>Report</td><td>D</td><td>Writing the report</td><td>2 hours</td><td>High degree of supervision and control</td></tr>
</table>

Outline of the stages in the Assignment

1 Research stage

Task A

Selecting a topic and a title

Although you can choose a topic yourself, your teacher will probably support you by providing some ideas to choose from. Make sure you select something relevant, that you are interested in and understand. It might be a good idea to think about using one of the experimental techniques that are needed for the Course. The table below reminds you about the techniques and, in each case, gives some possible related Assignment topics. **You should note that these are topics, not Assignment aims.**

Your title should be informative and might indicate the topic area, but your aim will need to be more specific.

Technique	Page	Assignment topic areas
Using gel electrophoresis	9	Comparison of proteins from different animal tissues
Measuring the effect of substrate concentration on enzyme activity	57	Effects of different substrate concentrations on the rate of catalase activity
Measuring the effect of inhibitor concentration on enzyme activity	57	Effects of inhibitor concentrations on catalase activity
Using a respirometer	65	Effects of different respiratory substrates on the respiration rates of yeast
Measuring metabolic rate using oxygen, carbon dioxide and temperature probes	65	Effect of temperature on the rates of fermentation in yeast
Using a spectroscope	106	Comparison of the of the absorption spectra of pigment extracts from leaves of different colours
Using paper or thin layer chromatography to separate pigments	106	Comparison of the ranges of pigments found in the leaves of sun and shade plants

There are many other possible experimental techniques you could use, leading to many other Assignment topics.

Devising an aim

Ensure that you have a clear and specific aim for the Assignment. For example, an acceptable aim might be 'To investigate the effect of substrate concentration on enzyme activity'. You will need to state your experimental aim clearly in your Report.

Summarising relevant background biology

The background biology which you give in your Report must be relevant to your aim. To score full marks, you must give at least four relevant points. Ideally, you should present your underlying biology material in a separate section of your Report. It is not necessary to give references for this section.

Planning your investigation

Think about what you will do. What apparatus will you need? What steps will you take? What measurements will you make? How will you collect and record your results data? You need to think about controls and reliability. You will have to summarise your methods in your Report.

Task B

Carrying out experiments or fieldwork

You must carry out experiments or fieldwork safely to generate data relevant to your aim. This will involve carrying out your planned experiment under the supervision of your teacher. You are able to work with a partner or small group and share results data.

Collecting numerical data

You must record your experimental data in numerical form. It is sensible to design and draw some table grids to take your results before you start carrying out your work. You must have your own copy of the raw results data. During your Report write up, you must process your data by, for example, calculating mean values and presenting the data as a line graph or bar chart, as appropriate to the data itself.

Drawing a conclusion

You must draw a conclusion. This has to be relevant to your aim and must be supported by the data and other information you have collected. If the pre-existing source of data does not support your own data, you must state this in your Report and say why you think this is. Note that the mark for concluding cannot be awarded if you have not stated an aim in the first place.

Evaluating

In your Report, you must evaluate any experiments or fieldwork you carried out. The evaluation can focus on reliability, validity and accuracy of measurements. You should suggest an improvement which could minimise the effects of any error.

Task C

Selecting information from existing sources

You must research your aim and select information from other sources such as books, journals or the internet. Your teacher will probably supply you with a number of sources from which you can choose.

Comparing your own data with an existing source

The information you find has to be compared with the data you generated during your experiments or fieldwork. If a comparison is not possible, you must say so in your Report and give an explanation of why not.

You must give a reference to the source of your pre-existing data in your Report, as shown in the table below.

Source	Reference
Website	Full URL for the page or pages, i.e. the URL www.bbc.co.uk is not acceptable but http://www.bbc.co.uk/education/guides/z46cwmn/revision is an acceptable reference
Journal	Title, author, journal title, volume and page number
Book	Title, author, page number and either edition or ISBN

2 Report stage

Task D

Writing up your Assignment Report

You will have to write up your Report under controlled conditions, with access only to notes. The maximum time allowed for this stage is 2 hours. Your Report should have the sections or features shown in the table below. The table also shows the allocation of marks.

You must not have a draft report or a draft of any section with you to work from. It is expected that you will write your Report with the assistance only of your notes.

Section	Checkpoint	✓	Marks
Aim	My aim clearly describes the purpose of my investigation		1
Underlying biology	I have given an account of the biology relevant to the aim of my investigation		4
Data collection and handling	I have given a brief account of the method used to collect experimental data		1
	I have included sufficient raw data from my experiment in my Report		1
	My data have been organised into a correctly produced table with means where appropriate		1
	I have included relevant data from the internet or from literature		1
	I have given a correct reference for the data from the internet or from literature		1
Graphical presentation	I have presented my data as a line graph or bar graph as appropriate		1
	My graph has suitable scales, labels and units		2
	My data have been plotted accurately with a line or clear bar tops as appropriate		1
Analysis	I have correctly compared my experimental data to the data collected from the internet or from literature OR I have correctly completed a calculation based on my experimental data and linked to my aim		1
Conclusion	I have stated a conclusion that is directly linked to my aim, and supported by my experimental data and internet or literature source		1
Evaluation	I have included an evaluation of my work		3
Structure	My Report has an informative title and is clear and concise		1
Total			20

Your exam

General exam revision: 20 top tips

These are very general tips and would apply to all your exams.

1 **Start revising in good time.**

Don't leave it until the last minute – this will make you panic and it will be impossible to learn. Make a revision timetable that counts down the weeks to go.

2 **Work to a study plan.**

Set up sessions of work spread through the weeks ahead. Make sure each session has a focus and a clear purpose. What will you study, when and why?

3 **Make sure you know exactly when your exams are.**

Get your exam dates from the SQA website and use the timetable builder tool to make up your own exam timetable. You will also get a personalised timetable from your school but this might not be until close to the exam period.

4 **Make sure that you know the topics that make up each Course.**

Studying is easier if material is in chunks – why not use the SQA chunks? Ask your teacher for help on this if you are not sure.

5 **Break the chunks up into even smaller bits.**

The small chunks should be easier to cope with. Remember that they fit together to make larger ideas. Even the process of chunking down will help!

6 **Ask yourself these key questions for each Course.**

Are all topics compulsory or are there choices? Which topics seem to come up time and time again? Which topics are your strongest and which are your weakest?

7 **Make sure you know what to expect in the exam.**

How is the paper structured? How much time is there for each question? What types of question are involved – multiple choice, restricted response, extended response?

8 **There is no substitute for past papers – they are simply essential!**

Papers for all Courses are on the SQA website – look for the past paper finder and download as PDF files. There are answers and mark schemes too. There is also a selected range available through Hodder Gibson.

9 Use study methods that work well for you.

People study and learn in different ways. Reading and looking at diagrams suits some people. Others prefer to listen and hear material – what about reading out loud or getting a friend or family member to do this for you? You could also record and play back material.

10 There are only three ways to put material into your long-term memory:

- practice – for example, rehearsal, repeating
- organisation – for example, making drawings, lists, diagrams, tables, memory aids
- elaborating – for example, winding the material into a story or an imagined journey

11 Learn actively.

Most people prefer to learn actively – for example, making notes, highlighting, redrawing and redrafting, making up memory aids, writing past paper answers.

12 Be an expert.

Be sure to have a few areas in which you feel you are an expert. This often works because at least some of them will come up, which can boost confidence.

13 Try some visual methods.

Use symbols, diagrams, charts, flash cards, post-it notes, etc. The brain takes in chunked images more easily than loads of text.

14 Remember – practice makes perfect.

Work on difficult areas again and again. Look and read – then test yourself. You cannot do this too much.

15 Try past papers against the clock.

Practise writing answers in a set time. As a rough guide, you should be able to score a mark per minute.

16 Collaborate with friends.

Test each other and talk about the material – this can really help. Two brains are better than one! It is amazing how talking about a problem can help you solve it.

17 Know your weaknesses.

Ask your teacher for help to identify what you don't know. If you are having trouble, it is probably with a difficult topic so your teacher will already be aware of this – most students will find it tough.

18 Have your materials organised and ready.

Know what is needed for each exam. Do you need a calculator or a ruler? Should you have pencils as well as pens? Will you need water or paper tissues?

19 Make full use of school resources.

Are there study classes available? Is the library open? When is the best time to ask for extra help? Can you borrow textbooks, study guides, past papers, etc.? Is school open for Easter revision?

20 Keep fit and healthy!

Mix study with relaxation, drink plenty of water, eat sensibly, and get fresh air and exercise – all these things will help more than you could imagine. If you are tired, sluggish or dehydrated, it is difficult to see how concentration is even possible.

Higher Biology: 20 top exam tips

These tips apply specifically to Higher Biology. Remember that your Assignment is worth 20% of your grade – the other 80% comes from the 120 marks in the examination. The examination consists of two separate papers.

Paper 1: multiple choice (25 marks)

1 You have 40 minutes for this paper. Answer on the grid provided.
2 *Don't leave blanks* – complete the grid for each question as you work through.
3 Try to answer each question in your head *without* looking at the options. If your answer is there, you are home and dry!
4 If you are not certain, choose the answer that seemed most attractive on *first* reading the answer options.
5 If you are forced to guess, try to eliminate options before making your guess. If you can eliminate three, you are left with the correct answer even if you don't recognise it!

Paper 2: short answer and extended response (95 marks)

6 You have 2 hours and 20 minutes for this paper. Answer in the spaces on the question paper.
7 Try to write neatly and keep your answers on the support lines if possible – these are designed to take the length of answer required.
8 Another clue to answer length is the mark allocation. Most questions have 1 mark and the answer can be quite short; if there are more marks available, your answer will need to be extended and may well have two, three or four parts.
9 The questions are usually laid out to reflect the sequence of the Course but remember that some questions are *designed* to cover more than one area.
10 There will be two or three extended response questions that are worth between 3 and 10 marks – these are marked by awarding 1 mark for each relevant point given in your answer.

11 There will be a data question and an experimental question, each worth between 5 and 8 marks. It is essential to read these question stems carefully – there will be information contained there which is needed to answer the questions fully.
12 Grade C (less demanding) questions usually start with 'State', 'Give' or 'Name'; Grade A (more demanding) questions begin with 'Explain', 'Describe' and 'Suggest' and are likely to have more than one part to the full answer.
13 Using abbreviations like DNA, ATP, 3PG and RuBisCO is fine.
14 Don't worry that some questions are about unfamiliar contexts. This is deliberate. Just keep calm and read the question carefully. You should be able to work out the answer from contexts you do know about and from information given in the question.
15 If a question contains a choice, be sure to spend enough time making the right choice.
16 Remember to *use values from the graph* when describing graph trends if you are asked to do this.
17 Draw graphs using a ruler and use the data table headings and units for the axes labels.
18 Look out for graphs with two *y*-axes – it is easy to make mistakes in their interpretation. A couple of highlighter pens might help to link data to axes.
19 Answers to calculations don't usually have more than two decimal places. If there is a *space for a calculation*, it is very likely that you will need to use it.
20 Don't leave blanks. Have a go, using the language in the question if you can.

Glossary

3'–5' (1.1) refers to the direction of a DNA strand with deoxyribose at the 3' end and phosphate at the 5' end

3-phosphoglycerate (3PG) (3.1b) produced when CO_2 is fixed to RuBP by RuBisCo

Absorption spectrum (3.1b) graph showing wavelengths of light absorbed by a pigment

Accuracy (4) the extent to which a measurement is near to its true value

Acetyl group (2.2) produced by breakdown of pyruvate; joins with oxaloacetate in the citric acid cycle

Action spectrum (3.1b) the rate of photosynthesis carried out at each wavelength of light

Activation energy (2.1b) input of energy required to start a chemical reaction

Active site (2.1b) region on an enzyme molecule where the substrate binds

Adenine (A) (1.1) base that pairs with thymine in DNA

Aestivation (2.5) type of dormancy that allows animals to survive in periods of high temperature or drought

Affinity (2.1b) the degree to which molecules are likely to combine with each other

Agriculture (3.1a) human practice of growing crops and keeping livestock to maintain food security

Alliance (3.5 and 3.6) link between individuals in primate social groups, often used to increase social status within the group

Allopatric speciation (1.7) speciation in which gene flow is prevented by a geographical barrier

Alternative RNA splicing (1.3) synthesis of different mature transcripts from the same primary transcript

Altruistic behaviour (3.5 and 3.6) behaviour which harms a donor but benefits the recipient

Amino acid (1.3) unit of polypeptide structure

Anabolic (2.1a) metabolic activity that requires energy input and builds up larger molecules from smaller ones

Animal welfare (3.3 and 3.4) relating to activities that are designed to be humane to livestock while maximising their yield

Annual weed (3.3 and 3.4) a weed that completes its life cycle in one year, which has rapid growth, high seed output and long-term seed viability

Anticodon (1.3) sequence of three bases on tRNA that specifies an amino acid

Antiparallel (1.1) parallel strands in DNA, running in opposite directions

Appeasement behaviour (3.5 and 3.6) behaviour carried out by subordinate animals to appease a dominant individual and reduce conflict

Archaea (1.8 and 2.6) a domain of life containing single-celled microorganisms

Artificial chromosome (2.7) used as a vector to carry foreign genetic information into another cell

ATP (2.2) molecule used for energy transfer in cells

ATP synthase (2.2) enzyme within a phospholipid membrane which produces ATP from ADP and Pi

Atria (2.3) chambers of a vertebrate heart that receive blood

Attachment site (1.3) position on a tRNA molecule at which a specific amino acid binds

Bacteria (1.8) a domain of life

Balance (4) instrument for measuring the weight or mass of materials

Base (1.1) a coding component of a nucleotide

Beaker (4) container to hold and pour liquids for use in experiments

Behavioural barrier (1.7) barrier to gene flow caused by behavioural differences between individuals

Bioaccumulation (3.3 and 3.4) build-up of a chemical in an organism

Biodiversity (3.7) variety and relative abundance of species

Bioinformatics (1.8) use of computers and statistics in analysis of sequence data

Biological control (3.3 and 3.4) method of controlling pests using natural predators, parasites or pathogen of the pest

Biological species (1.7) group of similar organisms interbreeding to produce fertile young

Biomagnification (3.3 and 3.4) increase in the concentration of a chemical between trophic levels

Biosynthesis (2.6) a multi-step, enzyme-catalysed process in living organisms in which substrates are converted into more complex products

Bottleneck effect (3.8) inability of a species to evolve due to lack of genetic diversity

Bt toxin gene (3.2) a gene that is inserted into plants using recombinant DNA technology to produce a protein that acts as a pesticide

Calorimeter (2.3) device for measuring heat production by organisms

Carbon dioxide and oxygen probes (2.3) devices for measuring the production of carbon dioxide and the uptake of oxygen by organisms

Carbon fixation stage (3.1b) Calvin cycle; second stage in photosynthesis, which results in the production of glucose

Carotenoids (3.1b) orange and yellow accessory pigments in plants that extend the range of wavelengths absorbed and pass the energy to chlorophyll for photosynthesis

Catabolic (2.1a) metabolic activity that releases energy from reactions which break down large molecules into smaller ones

Cellular differentiation (1.4) changes to cells involving switching on certain genes and switching off others

Cellular respiration (2.2) release of energy from respiratory substrates

Cellulose (3.1b) structural carbohydrate in cell walls derived from photosynthesis

Chlorophyll (3.1b) green pigment molecule in plants that absorbs red and blue light for photosynthesis

Chloroplast (1.1) organelle in which the chemical reactions of photosynthesis occur

Chromosome (1.1) structure that contains the genetic material of an organism encoded into DNA

Citrate (2.2) citric acid; first substance produced in the citric acid cycle

Citric acid cycle (2.2) second stage of aerobic respiration, occurring in the matrix of mitochondria

Codon (1.3) sequence of three bases on mRNA that specifies an amino acid

Coenzyme A (2.2) substance that carries an acetyl group into the citric acid cycle as acetyl coenzyme A

Coenzyme NADP (3.1b) hydrogen carrier in photosynthesis

Colorimeter (4) device used to measure the concentration of a coloured solution

Common ancestor (1.8) species from which more than one more modern species have diverged

Competition (3.1a) struggle for existence between two organisms

Competitive inhibition (2.1b) the slowing of reaction rate due to the presence of a substance resembling the substrate

Complete double circulation (2.3) double circulation with complete separation of oxygenated and deoxygenated blood (e.g. in birds and mammals)

Conformer (2.4) animal whose internal environment is dependent on its environment

Consequential dormancy (2.5) dormancy that occurs in response to the onset of adverse conditions

Control (4) technique which can help to ensure that experimental results are caused by the independent variable

Cooperative hunting (3.5 and 3.6) hunting behaviour in which individuals work together to catch prey

Crop plant pests (3.3 and 3.4) organisms such as insects, nematode worms and molluscs which reduce the yield of crops

Cross-breed population (3.2) population showing improved characteristics, produced by crossing individuals from different breeds

Cultivar (3.1a) varieties of cultivated crops, for example high yielding, disease resistant, GM cultivars

Cultural (3.3 and 3.4) method of crop protection based on human behaviours and activities such as ploughing, weeding and crop rotation

Cytosine (C) (1.1) base that pairs with guanine in DNA

Death phase (2.6) phase of microorganism growth in which death rate of cells exceeds rate of cell division

Dehydrogenase (2.2) enzyme which removes hydrogen from its substrate; important in the citric acid cycle

Deleterious sequence (1.7) DNA sequence that lowers survival rate

Deletion (1.5 and 1.6) chromosome mutation in which a sequence of genes is lost from a chromosome

Deletion of nucleotides (1.5 and 1.6) single gene mutation involving removal of a nucleotide from a sequence

Deoxyribose (1.1) pentose sugar that is a component of a DNA nucleotide

Dependent variable (4) variable which forms the results of an experiment

Directional selection (1.7) natural selection that favours one extreme of a phenotype

Disease forecasting (3.3 and 3.4) management system used to predict the occurrence of plant diseases; applications of fungicide based on disease forecasts are more effective than treating diseased crops

Disruptive selection (1.7) natural selection that favours two different phenotypes

DNA (1.1) deoxyribonucleic acid; a molecule that holds the genetic code in living organisms

DNA polymerase (1.2) enzyme that adds free complementary DNA nucleotides during replication of DNA

Domains of life (1.8) three major groups, Bacteria, Archaea and Eukarya, into which all known living species can be divided

Dominant (3.5 and 3.6) animal ranked at the top of a social hierarchy

Dominant species (3.7) most abundant species in an ecosystem

Donor (3.5 and 3.6) organism that carries out the altruistic behaviour to benefit the recipient

Dormancy (2.5) reduction in metabolic rate made by organisms to tolerate adverse conditions (e.g. hibernation, aestivation)

Double circulation (2.3) blood flows through the heart twice during a full circulation of the body

Double helix (1.1) three-dimensional shape of a DNA molecule

Dropping pipette/syringe (4) instrument for delivering small measured volumes of liquid

Duplication (1.5 and 1.6) chromosome mutation in which a section of chromosome is added from its homologous partner

Ecological barrier (1.7) barrier to gene flow caused by ecological preference differences between individuals

Ecological niche (2.4) the way of life and the role of an organism in its community

Ecosystem diversity (3.7) the number of distinct ecosystems within a defined area

Electron (2.2) negatively charged particle that yields energy as it passes through an electron transport chain

Electron transport chain (2.2) group of proteins embedded in membranes of mitochondria and chloroplasts

Embryonic stem cell (1.4) stem cell from an embryo that can divide and become any type of cell

End product (2.1b) a final product in a metabolic reaction which is not further converted to other substances

Energy investment phase (2.2) stage of glycolysis in which ATP is used up

Energy pay-off stage (2.2) stage of glycolysis in which a net gain of ATP is made

Environmental factors (1.3) these include light, temperature, nutrients and other factors which can affect the phenotype of an organism

Eukaryotic (1.1) refers to a domain of life that is characterised by cells with a discrete nucleus

Evolution (1.7) changes to organisms over time that are mainly caused by natural selection

Exon (1.3) sequence of DNA that codes for part of a protein

Exponential growth (2.6) growth phase of microorganisms involving a rapid geometric increase in numbers

F_1 (3.2) first generation of offspring from a genetic cross

F_2 (3.2) offspring of an F_1 generation

Feedback inhibition (2.1b) enzyme inhibition caused by the presence of an end product of a metabolic pathway acting as an inhibitor of an enzyme earlier in the pathway

Fermentation (2.2) production of pyruvate in the absence of oxygen

Fermenters (2.6) vessels for growing large quantities of microorganisms under optimum conditions

Fertiliser (3.1a) chemical addition to soil to increase plant growth

Field trials (3.2) non-laboratory tests carried out in a range of environments to compare the performance of different cultivars or treatments and to evaluate GM crops

Food security (3.1a) the ability to access sufficient quantity and quality of food over a sustained period of time

Fossil evidence (1.8) information derived from the remains of extinct organisms

Frame-shift mutations (1.5 and 1.6) single gene mutations which cause all codons and all amino acids after the mutation to be changed

Fungicide (3.3 and 3.4) chemical substance which kills fungal pest species

Funnel (4) device for channelling liquids or gas bubbles in an experiment; can be used with a filter paper to separate mixtures of solid and liquid

Gel electrophoresis (1.2) technique used to separate macromolecules, such as DNA fragments of different sizes

Gene (1.3) DNA sequence which codes for a protein

Gene expression (1.3) transcription and translation

Generation time (2.6) time taken for a microorganism cell to divide

Genetic diversity (3.7) the number and frequency of all the alleles within a population

Genetic vector (2.7) a DNA molecule such as a plasmid or artificial chromosome used to carry foreign genetic information into another cell

Genome (1.5 and 1.6) total genetic material present in an organism

Genomic sequencing (1.8) procedure used to reveal the nucleotide sequence of a genome

Geographical barrier (1.7) physical barrier to gene flow, such as a mountain or river

Glyceraldehyde-3-phosphate (G3P) (3.1b) compound in the carbon fixation stage that can be converted to glucose or used to regenerate RuBP

Glycolysis (2.2) first stage in cellular respiration, which occurs in the cytoplasm

Glyphosate resistance gene (3.2) a gene that is inserted into plants using recombinant DNA technology to provide herbicide tolerance

GM crop (3.2) genetically modified crop which contains a gene from another species

Growth medium (2.6) substance which provides microorganisms with an energy source and raw materials for biosynthesis

Guanine (G) (1.1) base that pairs with cytosine in DNA

Habitat corridor (3.8) link that allows movement of animals between habitat islands or fragments, increasing access to food and choice of a mate

Habitat fragment (3.8) very small area of isolated habitat

Heat-tolerant DNA polymerase (1.2) enzyme from hot-spring bacteria, used in PCR

Heterozygous (3.2) having two different alleles of the same gene and so not true breeding

Hibernation (2.5) response of an animal to survive adverse conditions by reduction of metabolic rate, brought about by low temperatures and lack of food

Histones (1.1) proteins with which DNA is associated in linear chromosomes

Homeostasis (2.4) maintenance of a steady state or constant internal environment in the cells of a living organism

Homozygous (3.2) having two identical alleles of the same gene and so true breeding

Horizontal gene transfer (1.7) inheritance of genetic material within a generation

Host (3.5 and 3.6) organism that is harmed by the loss of energy and nutrients to a parasite

Hybrid vigour (3.2) the increase in such characteristics as size, growth rate, fertility, and yield of a hybrid organism over those of its parents

Hydrogen bond (1.1) weak chemical link joining complementary base pairs in DNA

Hydrogen ion (H^+) (2.2) a hydrogen atom which has lost an electron, leaving it positively charged

Hypothalamus (2.4) temperature-monitoring centre of the mammalian brain containing thermoreceptors which detect changes in blood temperature

Inbreeding (3.2) crossing organisms of the same or similar genotype for several generations until the population breeds true for the desired characteristics

Inbreeding depression (3.2) an increase in the frequency of individuals who are homozygous for recessive deleterious alleles, which lowers biological fitness

Incomplete double circulation (2.3) double circulation with some mixing of oxygenated and deoxygenated blood (e.g. in amphibians and some reptiles)

Independent variable (4) variable being investigated and so purposely varied

Induced fit (2.1b) change to an enzyme's active site brought about by its substrate

Innate (2.5) unlearned instinctive behaviour

Inorganic phosphate (Pi) (2.2) used to phosphorylate ADP

Insertion (1.5 and 1.6) single gene mutation in which an additional nucleotide is placed into a sequence

Integrated pest management (IPM) (3.3 and 3.4) a combination of chemical, biological and cultural means to control pests

Intermediate (2.2) substance in a metabolic pathway between the original substrate and the end product

Intermediate (secondary) host (3.5 and 3.6) organism involved in the life cycle of a parasite but that is separate from the main host

Introduced (non-native) species (3.8) species that humans have moved either intentionally or accidentally to a new geographic location

Intron (1.3) non-coding sequence of DNA within a gene

Invasive species (3.8) naturalised species that spreads rapidly and eliminates native species, therefore reducing species diversity

Inversion (1.5 and 1.6) chromosome mutation in which a set of genes rotates through 180°

Isolation (1.7) refers to prevention of gene flow between populations of a species

Kin selection (3.5 and 3.6) situation in which the donor benefits in terms of the increased chance of survival of shared genes in the recipient's offspring or future offspring

Lactate (2.2) produced by the anaerobic conversion of pyruvate in mammalian muscle cells

Lag phase (2.6) phase when microorganisms adjust to the conditions of the culture by inducing enzymes that metabolise the available substrates

Lagging strand (1.2) DNA strand that is replicated in fragments

Last universal ancestor (1.8) the most recent organism from which all other organisms have descended

Leading strand (1.2) DNA strand that is replicated continuously

Learned (2.5) behaviour of an individual organism not common to all members of its species and which is acquired by experience

Legume (3.1a) plant with seeds in a pod, such as the bean or pea

Ligase (1.2) enzyme that joins DNA fragments to make the lagging strand

Light energy (3.1b) radiant energy used in photosynthesis

Livestock (3.1a) agricultural animals

Log phase (2.6) exponential phase of microorganism growth

Matrix (2.2) central fluid-filled cavity of a mitochondrion

Mature messenger RNA (mRNA) (1.3) carries a copy of the DNA code to a ribosome

Mature transcript (1.3) alternative term for mature mRNA

Measuring cylinder (4) container to hold and measure volumes of liquid

Meristem (1.4) region in a plant in which mitosis occurs

Metabolic pathway (2.1a) enzyme-controlled sequence of chemical reactions in cells

Metabolic rate (2.3) rate of consumption of energy by an organism

Metabolism (2.1a) total of all metabolic pathways in an organism

Metabolite (2.6) a substance produced by metabolism or a substance necessary for a particular metabolic process

Migration (2.5) response by an organism to avoid adverse conditions by relocating

Misdirected behaviour (3.3 and 3.4) inappropriate use of normal behaviour, such as over-grooming of feathers by chickens

Missense (1.5 and 1.6) refers to a single gene mutation which results in one amino acid in a protein being changed for another

Mitochondria (1.1) cell organelles in which the aerobic stages of respiration occur (singular: mitochondrion)

Model organisms (1.8) species which have had their genomes sequenced and can be used as references

Molecular clock (1.8) graph that shows differences in sequence data for a protein against time

Molecular interactions (1.3) various chemical links joining amino acids and giving protein molecules their shape

Mutagenesis (2.7) the process of inducing mutations by exposure to mutagenic agents such as ultraviolet (UV) light, other radiation, or certain chemicals

Multipotent (1.4) refers to stem cells which can differentiate into all types of cell found in a particular tissue

Mutations (1.5 and 1.6) random changes to DNA sequences

Mutualism (3.5 and 3.6) symbiosis in which both partners benefit from the relationship

NAD (2.2) coenzyme which carries hydrogen and electrons from glycolysis and the citric acid cycle to the electron transport chain

Native species (3.8) (indigenous) species occurring naturally in its ecosystem, having evolved there

Natural selection (1.7) process that ensures survival of the fittest

Naturalised species (3.8) species that becomes established within a wild community

Negative feedback (2.4) control system for maintaining homeostasis in regulator organisms

Net gain (2.2) overall production of ATP is greater than ATP used up

Non-coding sequence (1.5 and 1.6) DNA sequence that does not encode protein

Non-competitive inhibition (2.1b) enzyme inhibition by a substance that permanently alters the active site of an enzyme

Nonsense (1.5 and 1.6) refers to a single gene mutation which results in premature stop codon

Nucleotide (1.1) component of DNA consisting of a deoxyribose sugar, a phosphate group and a base

Origin of replication (2.7) site which allows self-replication of a plasmid or artificial chromosome

Oxaloacetate (2.2) substance that combines with the acetyl group in the citric acid cycle to form citrate

Parasite (3.5 and 3.6) organism in the symbiotic relationship that benefits in terms of energy or nutrients with the host being harmed by the loss of these resources

Parental care (3.5 and 3.6) activities performed by parents that increase the survival chances of their young

Peptide bonds (1.3) strong chemical links which join amino acids in the primary structure of polypeptides

Perennial weed (3.3 and 3.4) weed plant with storage organs and vegetative reproduction which persists in the community by continuing to grow year after year

Persistent (3.3 and 3.4) unable to be broken down by enzymes

Personalised medicine (1.8) possible future development in which treatment is based on an individual's genome

Pest (3.1a) organisms such as insects, nematode worms and molluscs, which damage agriculture and reduce food security

Pesticides (3.3 and 3.4) these include herbicides to kill weeds, fungicides to control fungal diseases, insecticides to kill insect pests, molluscicides to kill mollusc pests and nematicides to kill nematode pests

Pharmacogenetics (1.8) the study of inherited differences affecting metabolic pathways which can affect individual therapeutic responses to, and side-effect of, drugs

Phenotype (1.3) outward appearance of an organism

Phosphate (1.1) component of a DNA nucleotide that forms part of the sugar–phosphate backbone

Phospholipid membrane (2.1a) membrane of a cell made from fluid phospholipid molecules and proteins

Phosphorylation (2.2) addition of phosphate to a substance

Photolysis (3.1b) light energy splits water into oxygen, which is evolved, and hydrogen, which is transferred to the coenzyme NADP

Photosynthesis (3.1a) production of carbohydrate by a plant using the energy of light

Phylogenetics (1.8) study of evolutionary relatedness of species

Pigment (3.1b) coloured substance that absorbs light for photosynthesis

Plasmid (1.1) circular loop of genetic material found in prokaryotic organisms and some yeasts

Pluripotent (1.4) refers to an embryonic stem cell which can differentiate into any of the cell types in an organism

Polymerase chain reaction (PCR) (1.2) method of amplifying sequences of DNA *in vitro*

Polypeptide (1.3) short strand of amino acids

Pore (2.1a) small gap in a membrane created by a channel-forming protein

Predictive dormancy (2.5) dormancy that occurs before the onset of adverse conditions, triggered by environmental cues such as temperature and photoperiod

Primary transcript (1.3) molecule made when DNA is transcribed

Primates (3.5 and 3.6) mammalian group that includes monkeys, apes and humans

Primer (1.2) short complementary strand of DNA

Product (2.1b) substance resulting from an enzyme-catalysed reaction

Prokaryotic (1.1) refers to the domains of life characterised by cells with no discrete nucleus

Protein (1.1) large molecule made up from a chain of amino acids linked by peptide bonds

Pump (2.1a) protein in a phospholipid membrane that carries substances across it by active transport

Pyruvate (2.2) end product of glycolysis

Randomisation (3.2) a methodology based on chance, used to eliminate bias when measuring treatment effects

Recessive (3.2) alleles which only show in the phenotype when they are in homozygous form

Recipient (3.5 and 3.6) organism that benefits from the altruistic behaviour of the donor

Reciprocal altruism (3.5 and 3.6) a selfless behaviour which is returned by the original recipient to the original donor

Recombinant DNA technology (2.7) involves the joining together of DNA molecules from two different species

Recombinant DNA technology (3.2) single genes for desirable characteristics can be inserted into the genomes of crop plants, creating genetically modified plants with improved characteristics

Reflection (3.1b) light that strikes a leaf passes away from its surface back to the atmosphere

Regulator (2.4) an animal that can control its internal environment and maintain homeostasis by using physiological mechanisms

Regulatory sequences (2.7) DNA sequences which control gene expression

Relative abundance (3.7) the numbers of a species compared with others in a community

Reliability (4) measure of confidence in the results of an experiment

Replicates (3.2) repeats of field trials to take account of variability within the sample

Replication (1.2) formation of identical copies of DNA molecules

Research use (1.4) use as a model for study or for testing of drugs to develop new treatments

Resistant stage (3.5 and 3.6) some parasites use resistant larvae and pupae, which can survive adverse environmental conditions until a new host comes in contact with them

Respirometer (2.3) device for measuring the oxygen consumption of organisms

Restriction endonuclease (2.7) enzyme which cuts target sequences of DNA from a chromosome or is used to open a plasmid

Restriction sites (2.7) target sequences of DNA where specific restriction endonucleases cut

Ribose sugar (1.3) sugar component of an RNA nucleotide

Ribosomal RNA (rRNA) (1.3) type of RNA that makes up ribosomes

Ribosome (1.3) site of protein synthesis; composed of rRNA and protein

Ribulose bisphosphate (RuBP) (3.1b) the carbon dioxide acceptor molecule in the carbon fixation stage

Ritualistic threat display (3.5 and 3.6) behaviour such as body posture, raised hackles and baring teeth, used instead of physical aggression

RNA (1.3) ribonucleic acid, which occurs in several forms in cells

RNA polymerase (1.3) enzyme involved in synthesis of primary transcripts from DNA

RNA splicing (1.3) joining of exons following the removal of introns from a primary transcript

Root crop (3.1a) a crop that is a root vegetable, for example carrot or sugar beet

RuBisCo (3.1b) the enzyme that fixes carbon dioxide by attaching it to RuBP

Safety genes (2.7) genes which are introduced that prevent the survival of the microorganism in an external environment

Secondary metabolite (2.6) substance, not associated with growth, produced during the stationary phase of a culture of microorganisms, for example antibiotics

Selectable markers (2.7) genes present in the vector that ensure only microorganisms which have taken up the vector grow in the presence of the selective agent, for example an antibiotic

Selection of treatments (3.2) choice of treatments in a field trial such that a valid comparison can be made

Selective herbicides (3.3 and 3.4) weedkillers that have a greater and targeted effect on certain plant species (broad-leaved weeds)

Semi-logarithmic graph (2.6) graph with one logarithmic scale and one linear scale; a logarithmic scale is a non-linear scale which is used when there is a large range of quantities

Sequence data (1.8) information concerning amino acid or nucleotide base sequences

Single circulatory system (2.3) blood flows through the heart once during a full circulation of the body (e.g. in fish)

Single gene mutations (1.5 and 1.6) mutations which involve the alteration of a sequence of DNA nucleotide

Social hierarchies (3.5 and 3.6) a rank order within a group of animals consisting of a dominant and subordinate members

Social insect (3.5 and 3.6) insect which lives in a complex social colony, such as bees, wasps and termites

Source of error (4) a factor or feature of experimental procedure which can lead to mistakes

Speciation (1.7) evolutionary process by which new species are formed

Species (1.7) group of organisms which interbreed to produce fertile offspring

Species diversity (3.7) the number of different species in an ecosystem (the species richness) and the proportion of each species in the ecosystem (the relative abundance)

Species richness (3.7) the number of different species in an ecosystem

Spectroscope (4) device which shows the spectrum of light coming from a source

Splice-site mutation (1.5 and 1.6) mutation at a point where coding and non-coding regions meet in a section of DNA

Stabilising selection (1.7) natural selection that favours average phenotypes and selects against extremes

Starch (3.1b) storage carbohydrate in plants

Start codon (1.3) triplet transcribed from DNA to a primary transcript indicating the start of the gene

Stationary phase (2.6) phase of microorganism growth during which secondary substances can be made

Stem cell (1.4) cell that can divide and then differentiate in animals

Stereotypy (3.3 and 3.4) repetitive actions such as aimless chewing movements in pigs

Sterility (2.6) not containing contaminating microorganisms

Stop codon (1.3) triplet on the primary transcript which signals a stop to translation

Stopwatch (4) instrument for measuring time in an experiment

Subordinate (3.5 and 3.6) an animal ranked below the dominant individual

Substitution (1.5 and 1.6) single gene mutation in which one nucleotide is replaced by another

Substrate (2.1b) substance on which an enzyme acts

Sugar–phosphate backbone (1.1) strongly bonded strand of DNA

Symbiosis (3.5 and 3.6) an intimate coevolved relationship between two different species

Sympatric speciation (1.7) speciation in which gene flow is prevented by ecological or reproductive barriers

Systemic herbicide (3.3 and 3.4) weedkiller that spreads through the vascular system (phloem) of the plant and prevents regrowth

Template strand (1.2) DNA strand on which a complementary copy is made

Test tube (4) glass container for holding liquid mixtures in experiments

Therapeutic use (1.4) used as part of medical therapy

Thermometer (4) instrument for measuring temperature in an experimental situation

Thermoreceptor (2.4) heat-sensitive cell in the hypothalamus of mammals

Thermoregulation (2.4) use of negative feedback in regulation of body temperature in mammals

Thymine (T) (1.1) base that pairs with adenine in DNA

Tissue (adult) stem cell (1.4) stem cell from a tissue that can divide and differentiate to become any of the cells in that tissue

Torpor (2.5) period of reduced activity in organisms with high metabolic rates such as small birds and mammals

Total cell count (2.6) total number of cells in a culture including viable (live) cells and dead cells

Transcription (1.3) copying of a DNA sequence to make a primary transcript

Transfer RNA (tRNA) (1.3) transfers specific amino acids to the mRNA on ribosomes

Translation (1.3) production of a polypeptide using sequences on mRNA

Translocation (1.5 and 1.6) mutation in which part of a chromosome becomes attached to a non-homologous chromosome

Transmission (3.1b) physical process of passing light energy through a surface

Trophic level (3.1a) feeding level in a food chain

Uracil (1.3) RNA base not found in DNA; complemetary to adenine in transcription

Validity (4) measure of the fairness of an experiment

Variable (4) factor in an experiment which can change

Vasoconstriction (2.4) narrowing of the blood vessels resulting from contraction of the muscular wall of the vessels

Vasodilation (2.4) widening of blood vessels that results from relaxation of the muscular walls of the vessels

Vector (3.5 and 3.6) an organism that carries and transmits the parasite into a new host organism

Vegetative reproduction (3.3 and 3.4) a form of asexual reproduction in plants

Ventricle (2.3) chamber of a vertebrate heart that distributes blood

Vertical gene transfer (1.7) inheritance of genetic material from parents by offspring

Viable cell count (2.6) number of live cells from a total cell count

Water bath (4) trough of thermostatically heated water to provide steady, controlled temperatures

Yeast (1.1) a special eukaryote which contains plasmids